OMNIDIRECTIONAL OPTICAL FILTERS

OMNIDIRECTIONAL OPTICAL FILTERS

by Vladimir Kochergin
Lake Shore Cryotronics, Inc.
Westerville, OH USA

Kluwer Academic Publishers
Boston/Dordrecht/London

Distributors for North, Central and South America:
Kluwer Academic Publishers
101 Philip Drive
Assinippi Park
Norwell, Massachusetts 02061 USA
Telephone (781) 871-6600
Fax (781) 871-6528
E-Mail <kluwer@wkap.com>

Distributors for all other countries:
Kluwer Academic Publishers Group
Post Office Box 322
3300 AH Dordrecht, THE NETHERLANDS
Telephone 31 78 6576 000
Fax 31 78 6576 474
E-Mail <orderdept@wkap.nl>

Electronic Services <http://www.wkap.nl>

Library of Congress Cataloging-in-Publication

Kochergin, Vladimir
Omnidirectional optical filters/by Vladimir Kochergin.
p.cm.
ISBN 978-1-4419-5342-1
1. Optical communications. 2.Digital filters (Mathematics) I. Title.
TK5103.59.K 63 2003
621.382'7--dc21

2002043386

Printed on acid-free paper.

Printed in the United States of America

CONTENTS

Acknowledgments

The material of this book was prepared with a great assistance of dozens of colleagues. I would like to thank Lake Shore Cryotronics, Inc. management team, Mr. Michael Swartz particularly, for the financial support of my research. I would like to thank Dr. Philip R. Swinehart for the discussions, directions, suggestions and encouragement during my research. I am also grateful to Professors Yang Zhao and Ivan Avrutsky, who introduced me to this field of research. I would like to thank Mr. Ralph F. Orban for the great help in silicon anodization experiments and Mr. Jeff Hardman for the assistance in reactive ion etching. I would like to thank Dr. Saman Dharmatilleke for the help in electroplating. I am grateful to Dr. Kimberly D. Pollard for the useful discussions of the chemistry of silicon anodic etching and Mr. Christopher McGee for the help with drawings. I would like to thank Professor Ivan Avrutsky for the help in optical evaluations of the macroporous filters. I am also grateful to Mr. Michael S. Hackett and Molly G. Maloof for the help in editing the text of this book.

Chapter 1

INTRODUCTION

Optical filters are among the most frequently used optical devices. Application areas include optical communications, imaging, spectroscopy, lithography, astronomy, and almost any other technology dealing with light. Various optical filters have been developed and differ by optical properties, applications, physical base, and so on. However, all optical filters modify of the transmission or reflection spectra.

Optical filter development can be related to color glass fabrication, which has been known for thousands of years. Specific impurities caused the coloring of the otherwise transparent glass. Beautiful windows and other pieces of art have been created using this technique. The next big step forward in optical filter design was not made until the first half of the twentieth century, when thin-film optical filters were introduced ([1] to [3]). The fast expansion of thin-film filters was in large part due to advancements in thin-film coating processes.

Thousands of years after the first colored glass was fabricated, optical filter development is still a hot topic. A big boost for optical filter development was the introduction of optical communications in the last quarter of the twentieth century. In optical communications optical filters play an important role in wavelength multiplexing and demultiplexing of optical signals. New types of optical filters have been developed for these purposes – waveguide and in-fiber optical filters (fiber or waveguide Bragg gratings [4], arrayed-waveguides gratings [5]).

Separately, artificial optical materials have been introduced, allowing a high degree of control of the optical properties of these materials. The classic examples of artificial optical materials are *photonic crystals* (see, for example [6]). The dielectric properties of photonic crystal (or *photonic bandgap material*, PBG) are spatially modulated with the period compared with the wavelength of light, leading to the appearance of photonic band gaps, preventing light from propagation in certain directions at specified wavelengths. It is the generalization of a one-dimensional periodic multilayer (like that used in dielectric mirrors and other classical thin-film filters) to second and third dimensions. Three-dimensional photonic crystals with a complete photonic band gap (a range of frequencies where the light of any polarization state at any angle of incidence is permitted to propagate through it) are potential candidates for various optical filters (reflectors, band-pass, or band-blocking filters- for example, [7]). They offer omnidirectionality of the filtering characteristics. In fact, consideration of one-dimensional multilayers from the PBG point of view led to the realization of omnidirectional dielectric mirrors [8], which will be considered in more detail in Chapter 4 of this book. Two-dimensional PBG structures were also shown to be quite promising waveguides filters (for example, [9] to [13]) although they are based on completely different principles.

Another type of artificial optical material used as an optical filter was introduced in the last quarter of the twentieth century – arrays of metallic waveguides ([14] to [16]). This type of filter was introduced first for the far infrared spectral range (due to manufacturing difficulties for shorter wavelengths). However, recently a similar approach was successfully applied to deep ultraviolet filters [17]. Generalization of such a structure (examined in Chapters 5 and 6 of this book) also promises near-omnidirectional behavior in a reasonably wide angular range.

The aim of this book is to introduce the reader to new optical filter design concepts that offer the suppression of the angular dependence of the filtering characteristics. The book does not pretend to provide a thorough description of the state of art in optical filter design. Instead, three main types of omnidirectional filters (reflectors, band-pass, and short-pass) are addressed. A theoretical discussion is given for all three concepts, while the fabrication process description is given for only two of them, where manufacturing steps are dramatically different from prior optical filters.

This text is designed for a broad audience. No prior knowledge of optical concepts is required. But readers, who are familiar with optics will find this book considerably simpler to read and understand than beginners will. Interested undergraduates should find the contents approachable, while professionals (engineers and researchers) can find some interesting materials and perspectives that are not available elsewhere.

The text begins with a brief description of the basic optical concepts needed to understand the content of the remaining chapters. It starts from the Maxwell equations and propagation of plane electromagnetic waves (the cornerstone of any theoretical description of optical phenomena). The boundary conditions are briefly explained and applied to the analyses of reflection, transmission, and absorption of plane electromagnetic waves on the single-boundary, single-layer, and multilayer structure. This is the basis of all further descriptions in the book and inexperienced readers should carefully address it.

Next, enough optical waveguide theory is given to understand Chapters 5 and 6. Planar and cylindrical optical waveguides as well as planar leaky waveguides are discussed. At the end, Bragg grating and codirectional coupling phenomena in optical waveguides are discussed from the viewpoint of coupling-mode theory.

A brief description of the classical types of optical filters is given in Chapter 3. Quarter-wave stack and Fabry-Perot interferometer are discussed. Based on these phenomena, a brief theoretical description is given for metallic and dielectric mirrors, absorption-based and thin-film edge filters, and band-pass and narrowband-pass filters. The disadvantages of classic filter design concepts are noted. In Chapter 4 the omnidirectional dielectric reflector (concept, introduced recently by [8]) is discussed from a theoretical point of view. A Bloch-wave approach for multilayer media analyses is given in a very short form, and introduces band structure concept. Enlargement of the omnidirectional reflection zone of dielectric multilayer is discussed.

Short-pass ultraviolet, bandpass visible and infrared filters based on an array of waveguides are discussed in Chapters 5 and 6 respectively. Thorough theoretical descriptions of such structures are given, since such filters are quite different from classical optical filters. Special attention is given to the method of manufacturing of waveguide (or leaky waveguide) arrays. In particular, a detailed look at macroporous silicon fabrication is presented.

Chapter 2

BASIC THEORY

This chapter is intended to introduce the reader who is new in the field to some to some of the basic optical concepts that are needed to understand the material that is presented in other chapters. It starts with a general description of electromagnetic wave propagation, multilayer transmission, reflection, and absorption. The experienced reader can skip it.

2.1 Maxwell's Equations and Plane Electromagnetic Waves

The electromagnetic field (which occurs under the presence of charges in space) is characterized by two vectors: ***E***, the electric field vector, and ***B***, the magnetic induction vector. The electromagnetic fields affect matter and vice versa. Different vectors are used to take the electromagnetic field-to-matter interaction into account: ***D***, the electric displacement vector, ***H***, the magnetic field vector, and ***j***, the displacement current. The equations connecting these vectors with each other are called Maxwell's equations:

$$\vec{\nabla}\times\vec{E}+\frac{1}{c}\cdot\frac{\partial\vec{B}}{\partial t}=0 \tag{2.1}$$

$$\vec{\nabla}\times\vec{H}-\frac{1}{c}\cdot\frac{\partial\vec{D}}{\partial t}=\frac{4\pi}{c}\cdot\vec{j} \tag{2.2}$$

$$\vec{\nabla}\cdot\vec{D}=4\pi\rho \tag{2.3}$$

$$\vec{\nabla}\cdot\vec{B}=0 \tag{2.4}$$

Another set of equations, material equations, includes the following:

$$\vec{j}=\sigma\vec{E} \tag{2.5}$$

$$\vec{D}=\varepsilon\vec{E} \tag{2.6}$$

$$\vec{B}=\mu\vec{H}, \tag{2.7}$$

where σ is electric conductivity, ε is dielectric permittivity, μ is magnetic permeability, ρ is electric charge density, and c is speed of light in a vacuum. For isotropic media σ, ε, and μ are scalars, while for anisotropic media they are tensors. In most cases dielectric permittivity and magnetic permeability can be considered to be independent on the electromagnetic field strength.

However, for almost any medium, this approximation is not sufficient for strong electromagnetic fields. The strength of the electromagnetic field, starting from which the dielectric permittivity and/or magnetic permeability of the medium cannot be considered as independent of the field strength, depends on both the medium properties and the needed accuracy of approximation. For most optical filter applications this nonlinearity of medium properties can be neglected and, hence, it is beyond of the scope of this book. In the following discussion both dielectric permittivity and magnetic permeability will be considered to be linear.

Equations (2.1) to (2.7) determine both the electromagnetic field and, through material equations, the interaction of the electromagnetic field with the medium. Equations (2.1) to (2.7) have different solutions for different σ, ε, and μ. In optics both the charge density and the current density inside the medium can be considered to be zero, i.e., $\rho = 0$ and $\boldsymbol{J} = 0$. Under these conditions Maxwell's equations (2.1) to (2.7) can be transformed into so-called *wave equations* (for mathematical derivation see, for example, [18]):

$$\nabla^2 \vec{E} - \frac{\varepsilon\mu}{c^2} \cdot \frac{\partial \vec{E}}{\partial t} = 0 \tag{2.8}$$

$$\nabla^2 \vec{H} - \frac{\varepsilon\mu}{c^2} \cdot \frac{\partial \vec{H}}{\partial t} = 0 \tag{2.9}$$

Equations (2.8) and (2.9) are called the *wave equations*. The solutions of (2.8) and (2.9) include the electromagnetic monochromatic plane waves:

$$\psi = A \cdot e^{i(\omega \cdot t - \vec{k} \bullet \vec{r})} \tag{2.11}$$

where A is a constant, having a meaning of amplitude of the wave, ω is an angular frequency of the wave, and $\boldsymbol{k}$ is the wave vector. ψ can represent any Cartesian component of $\boldsymbol{E}$ and $\boldsymbol{H}$. Such a wave propagates through the medium with the velocity

$$v = \frac{\omega}{\left|\vec{k}\right|} = \frac{c}{\sqrt{\varepsilon\mu}} \tag{2.12}$$

The electromagnetic wave defined by (2.11) is called a *plane wave* because the surfaces of the constant phase (often called *wavefronts*) of such a wave are planes, normal to the wave vector $\boldsymbol{k}$ at any time t. The velocity (2.12) is called the *phase velocity* of the electromagnetic wave. The value $n = \sqrt{\varepsilon\mu}$ is the *index of refraction* or *refractive index* of the medium.

As mentioned above, ***E*** and ***H*** are vectors. Maxwell equations (2.1) to (2.7) set the limitations on the possible orientation of ***E***, ***H***, and ***k*** in a plane electromagnetic wave. If, according to (2.11), the fields of such a wave will be written as

$$\vec{E} = \vec{e} \cdot E_0 \cdot e^{i(\omega \cdot t - \vec{k} \bullet \vec{r})} \tag{2.13}$$

$$\vec{H} = \vec{h} \cdot H_0 \cdot e^{i(\omega \cdot t - \vec{k} \bullet \vec{r})} \tag{2.14}$$

then the following will be true:

$$\boldsymbol{e} \cdot \boldsymbol{k} = \boldsymbol{h} \cdot \boldsymbol{k} = 0; \;\; \boldsymbol{h} = (\, \boldsymbol{k} \times \boldsymbol{e}\,) / |\boldsymbol{k}|; \; H_0 = (\varepsilon / \mu)^{1/2} E_0 \tag{2.15}$$

E and ***H*** are both perpendicular to the direction of propagation ***k***/|***k***|, or, in other words, electromagnetic waves are transverse waves. Moreover, according to (2.15), ***E*** and ***H*** are also orthogonal to each other and are in phase.

According to (2.13) to (2.15), the plane electromagnetic wave can be fully characterized by its angular frequency ω, direction of propagation ***k***/|***k***|, and the directions of oscillations of any of its field vectors, usually called *polarization states* of the electromagnetic wave. The polarization state of the electromagnetic wave is the important parameter in optics. To illustrate the meaning of it, let us suppose that the monochromatic plane electromagnetic wave is propagating in the z- direction so that the electric field vector lies in the xy plane. In this case the electric field projections on the x and y axis can be written as

$$E_x = E_0^x \cos(\omega t - k \cdot z + \delta_x) \quad E_y = E_0^y \cos(\omega t - k \cdot z + \delta_y) \tag{2.16}$$

By eliminating $\omega t - k \cdot z$ in (2.16) one can obtain

$$\left(\frac{E_x}{E_x^0}\right)^2 + \left(\frac{E_y}{E_y^0}\right)^2 - 2\frac{\cos\delta}{E_x^0 E_y^0} E_x E_y = \sin^2\delta \tag{2.17}$$

where $\delta = \delta_y - \delta_x$, $-\pi < \delta \leq \pi$. Equation (2.17) shows that in general the end point of the wave electric field vector ***E*** moves around an ellipse. However, when $\delta = \delta_y - \delta_x = m\pi$ $(m = 0, 1)$, the polarization ellipse degenerates into a straight line. Such a polarization state of a monochromatic plane electromagnetic wave is called *linear polarization*. Alternatively, when the condition $\delta = \delta_y - \delta_x = \pm\pi/2$ is satisfied, the polarization ellipse will reduce to a circle. Such a polarization state of a monochromatic plane electromagnetic wave is called *circular polarization*. For more details on this topic the reader can refer to the classical textbooks of [18] or [19].

2.2 Boundary Conditions

The Maxwell's equations and electromagnetic wave characteristics have been obtained so far only for regions in space in which the physical properties of matter (characterized by ε and μ) are continuous (i.e., in a single medium). In contrast, optics usually deals with situations when ε and μ are changed abruptly on one or more interfaces. The field vectors ***E***, ***H***, ***D***, and ***B*** on both sides of each interface are connected with each other according to certain laws, also known as *boundary laws*. Consider a boundary surface separating two media with different dielectric permittivity and magnetic permeability (medium 1 and medium 2). In this case the boundary conditions on this surface can be written in the following form

$$\boldsymbol{n} \cdot (\boldsymbol{B}_2 - \boldsymbol{B}_1) = 0 \qquad \boldsymbol{n} \cdot (\boldsymbol{D}_2 - \boldsymbol{D}_1) = \sigma \tag{2.18a}$$
$$\boldsymbol{n} \times (\boldsymbol{E}_2 - \boldsymbol{E}_1) = 0 \qquad \boldsymbol{n} \times (\boldsymbol{H}_2 - \boldsymbol{H}_1) = \boldsymbol{K} \tag{2.18b}$$

where ***n*** is the unit vector, normal to said boundary surface and directed from medium 1 into medium 2, σ is the surface charge density, and ***K*** is the surface current density. As mentioned above, in most cases in optics both σ and ***K*** could be considered to be zero. Hence, according to (2.18), the tangential components of ***E*** and ***H*** and the normal components of ***D*** and ***B*** are continuous across the interface separating media 1 and 2. Consequently, (2.18) can be rewritten in more common form:

$$B_{2n} = B_{1n}, \qquad D_{2n} = D_{1n} \tag{2.19a}$$
$$\boldsymbol{E}_{2t} = \boldsymbol{E}_{1t}, \qquad \boldsymbol{H}_{2t} = \boldsymbol{H}_{1t} \tag{2.19b}$$

where $B_{2n} = \boldsymbol{n} \cdot \boldsymbol{B}_2$, $B_{1n} = \boldsymbol{n} \cdot \boldsymbol{B}_1$, $D_{2n} = \boldsymbol{n} \cdot \boldsymbol{D}_2$, $D_{1n} = \boldsymbol{n} \cdot \boldsymbol{D}_1$, and the subscript t denotes tangential component of the field vector. Boundary conditions (2.19) are the basis of calculations for the electromagnetic waves in real optical systems, for example, multilayer optical filters, optical waveguides, and many more.

2.3 Reflectance and Transmittance on the Single Boundary

Let us start our consideration with the simple case of reflection and transmission of a monochromatic plane electromagnetic wave from the single boundary between two isotropic media. The reflection and transmission of a plane light wave through the plane interface between two different media are the basic phenomena of all optics. A plane electromagnetic wave incident on the plane interface (which will be called as a single boundary later in this discussion) will be split into two plane electromagnetic waves: the reflected wave propagating of this interface into the medium of incidence and the transmitted wave propagating forward into the second medium (see figure 2.1).

For simplicity of explanation let us assume that the incident wave is linearly polarized. Let the dielectric permittivity be ε_0, magnetic permeability be μ_0 for medium 1 (medium of incidence), and let them be ε_1 and μ_1 for medium 2. Let us call the wavevector of the incident wave $\boldsymbol{k}_i$, of the reflected wave $\boldsymbol{k}_r$ and of the transmitted wave $\boldsymbol{k}_t$. Since, as shown above, the electric field vector of the plane electromagnetic wave in general has the form $\boldsymbol{E} = \mathbf{e}\,A \exp\,[i(\omega\, t - \boldsymbol{k}\cdot\boldsymbol{r})]$, the incident, reflected, and transmitted waves would have the form $\mathbf{e}_i\,A_i \exp\,[i(\omega\, t - \boldsymbol{k}_i\cdot\boldsymbol{r})]$, $\mathbf{e}_r\,A_r \exp\,[i(\omega\, t - \boldsymbol{k}_r\cdot\boldsymbol{r})]$, and $\mathbf{e}_t\,A_t \exp\,[i(\omega\, t - \boldsymbol{k}_t\cdot\boldsymbol{r})]$. A_i, A_r, and A_t denote amplitudes of incident, reflected, and transmitted waves, while $\mathbf{e}_i$, $\mathbf{e}_r$, and $\mathbf{e}_t$ denote the electric field unit vectors of incident, reflected, and transmitted waves. At the boundary $z = 0$ the arguments of the field amplitude of incident, reflected, and transmitted waves must satisfy the following equation:

$$(\boldsymbol{k}_i\cdot\boldsymbol{r})_{z=0} = (\boldsymbol{k}_r\cdot\boldsymbol{r})_{z=0} = (\boldsymbol{k}_t\cdot\boldsymbol{r})_{z=0}. \qquad (2.20)$$

Since the refractive indices of media 1 and 2 will be $n_0=\sqrt{\varepsilon_0\mu_0}$ and $n_1=\sqrt{\varepsilon_1\mu_1}$, respectively, the wavevector magnitudes will be

$$|\,\boldsymbol{k}_i\,| = |\,\boldsymbol{k}_r\,| = (\omega/c)\,n_0, \qquad |\,\boldsymbol{k}_t\,| = (\omega/c)\,n_1 \qquad (2.21)$$

Equations (2.20) and (2.21) define the directions of the reflected and transmitted waves with respect to the incident wave and the plane boundary. One of the consequences of (2.20) and (2.21) is that all three wavevectors $\boldsymbol{k}_i$, $\boldsymbol{k}_r$, and $\boldsymbol{k}_t$ must lie in a single plane. This plane is called *plane of incidence*. Note that both media 1 and 2 are assumed to be isotropic. Otherwise this statement is not always true. Another consequence of (2.20) and (2.21) is that tangential components of $\boldsymbol{k}_i$, $\boldsymbol{k}_r$, and $\boldsymbol{k}_t$ must be equal to each other. Let us define the angles between the wavevector and the normal direction to the surface as θ_i, θ_r, and θ_t for incident, reflected, and transmitted waves respectively (see figure 2.1).

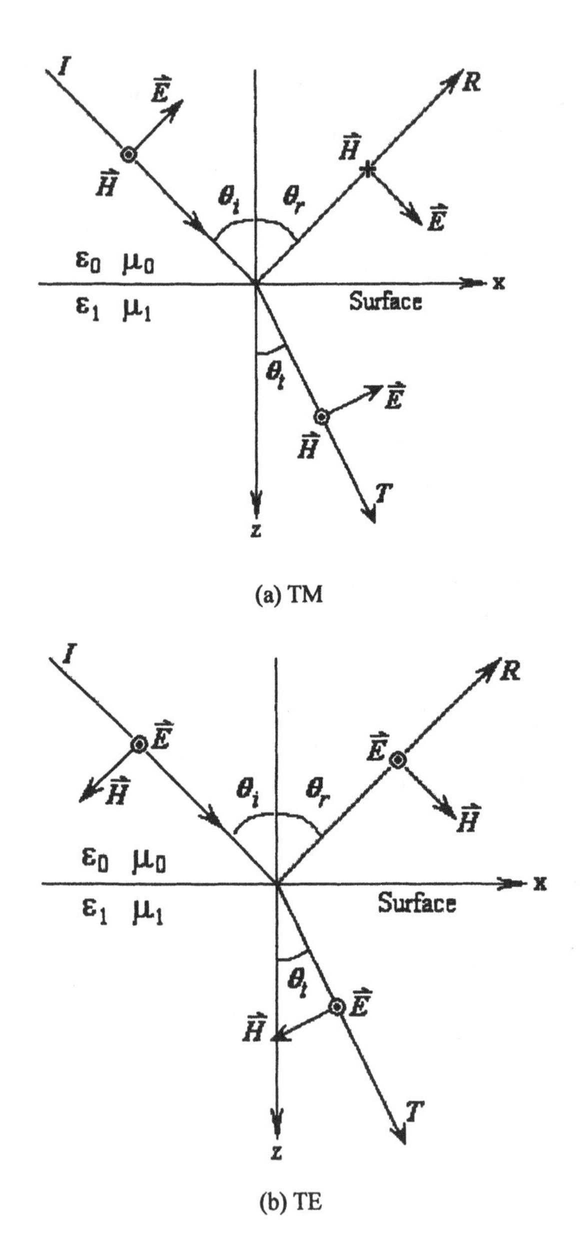

Figure 2.1. Electric and magnetic fields of incident on, reflected from and transmitted through the single boundary light waves for *p*-polarization (TM waves) (a) and *s*-polarization (TE waves) (b) cases.

In this case, according to (2.20) and (2.21),

$$n_0 \sin \theta_i = n_0 \sin \theta_r = n_1 \sin \theta_t. \tag{2.22}$$

This means that θ_i, θ_r, and θ_t are connected through Snell's law:

$$\frac{\sin\theta_i}{\sin\theta_t} = \frac{n_1}{n_0} \tag{2.23}$$

To obtain the relationship among A_r, A_t, and A_i one needs to solve boundary conditions (2.19). It is convenient to solve (2.19) separately for the field vector components parallel (denoted by p) and perpendicular (denoted by s) to the plane of incidence. p- and s-waves are independent from each other if both media 1 and 2 are homogeneous and isotropic. It means that for the s-polarized incident wave, both transmitted and reflected waves will be also s-polarized, while for the p -polarized incident wave, both transmitted and reflected waves will be also p -polarized. Another abbreviation for the s – polarized wave is TE wave, since the electric field vector $\boldsymbol{E}$ is transverse to the plane of incidence (figure 2.1b). Similarly, another abbreviation for the p – polarized wave is TM wave, since the magnetic field vector $\boldsymbol{H}$ is perpendicular to the plane of incidence (figure 2.1b). Below are the derivations of reflection and transmission coefficients, derived separately for TE and TM cases. The magnetic permittivity of both media is assumed to be unity ($\mu_1 = \mu_2 = 1$). Such approximation provides enough accuracy for the spectral ranges that are considered in this book.

2.3.1 TM Polarization

Let us introduce the x- and z-components of wavevectors in both media as k_x (since according to (2.20), x-components of wavevectors are the same in $\boldsymbol{k}_i$, $\boldsymbol{k}_r$ and $\boldsymbol{k}_t$) and k_{iz}, i = 0,1 (0 means first medium while 1 means second). Since the coordinate system (see figure 2.1a) is chosen that the plane of incidence coincides with the XOZ plane, k_{iy} = 0; i = 0,1. Hence, $k_{iz} = \sqrt{\varepsilon_i\left(\frac{\omega}{c}\right)^2 - k_x^2}$, i = 0,1. According to (2.15) for the TM-polarization case, the electric and magnetic field unit vectors can be written in the following form

$$\vec{e}_i = \left(1,0,-\frac{k_x}{k_{0z}}\right),\ \vec{e}_r = \left(1,0,\frac{k_x}{k_{0z}}\right),\ \vec{e}_t = \left(1,0,-\frac{k_x}{k_{1z}}\right),$$

$$\vec{h}_i = \frac{\omega}{c}\left(0, -\frac{\varepsilon_0}{k_{0z}}, 0\right),\ \vec{h}_r = \frac{\omega}{c}\left(0, \frac{\varepsilon_0}{k_{0z}}, 0\right),\ \vec{h}_t = \frac{\omega}{c}\left(0, -\frac{\varepsilon_1}{k_{1z}}, 0\right)$$

If we denote the amplitude of the electric field vector of the wave propagating in the positive direction in the media (i+ 1) by $E_+^{(i)}$, $i = 0,1$ and the amplitude of the electric field vector of the reflected wave by $E_-^{(0)}$, then the total electric field in medium 1, $\boldsymbol{E}^{(0)}$, will be the sum of the electric fields of incident and reflected waves:

$$\vec{E}^{(0)} = E_+^{(0)} \cdot e^{i \cdot k_{0z} z}\left(1, 0, -\frac{k_x}{k_{0z}}\right) + E_-^{(0)} \cdot e^{-i \cdot k_{0z} z}\left(1, 0, \frac{k_x}{k_{0z}}\right) \tag{2.24a}$$

Similarly, the total electric field in medium 2, $\boldsymbol{E}^{(1)}$, will be

$$\vec{E}^{(1)} = E_+^{(1)} \cdot e^{i \cdot k_{1z} z}\left(1, 0, -\frac{k_x}{k_{1z}}\right) \tag{2.24b}$$

The total magnetic field in the medium 1, $\boldsymbol{H}^{(0)}$, will be the sum of magnetic fields of incident and reflected waves:

$$\vec{H}^{(0)} = \frac{\omega}{c} E_+^{(0)} \cdot e^{i \cdot k_{0z} z}\left(0, -\frac{\varepsilon_o}{k_{0z}}, 0\right) + \frac{\omega}{c} E_-^{(0)} \cdot e^{-i \cdot k_{0z} z}\left(0, \frac{\varepsilon_o}{k_{0z}}, 0\right) \tag{2.24c}$$

The magnetic field in the medium 2, $\boldsymbol{H}^{(1)}$, will be

$$\vec{H}^{(1)} = \frac{\omega}{c} E_+^{(1)} \cdot e^{i \cdot k_{1z} z}\left(0, -\frac{\varepsilon_1}{k_{1z}}, 0\right) \tag{2.24d}$$

The boundary conditions (2.19) in this case will take the form

$$E_x^{(0)}\Big|_{z=0} = E_x^{(1)}\Big|_{z=0} \qquad H_y^{(0)}\Big|_{z=0} = H_y^{(1)}\Big|_{z=0} \tag{2.25}$$

By substituting (2.24) into (2.25) and using some elementary math one can find that

$$\frac{E_-^{(0)}}{E_+^{(0)}} = \frac{\dfrac{\varepsilon_o}{k_{0z}} - \dfrac{\varepsilon_1}{k_{1z}}}{\dfrac{\varepsilon_o}{k_{0z}} + \dfrac{\varepsilon_1}{k_{1z}}}, \qquad \frac{E_+^{(1)}}{E_+^{(0)}} = \frac{2\dfrac{\varepsilon_o}{k_{0z}}}{\dfrac{\varepsilon_o}{k_{0z}} + \dfrac{\varepsilon_1}{k_{1z}}} \tag{2.26}$$

Equation (2.26) defines the relationship among amplitudes of incident, reflected, and transmitted waves. The ratio of the amplitudes of the reflected wave and the incident wave is usually called complex *reflection coefficient*, r_p, while the ratio of the amplitudes of the transmitted wave and the incident wave is usually called complex *transmission coefficient*, t_p:

$$r_p = \frac{E_-^{(0)}}{E_+^{(0)}}, \qquad t_p = \frac{E_+^{(1)}}{E_+^{(0)}} \tag{2.27}$$

According to (2.27) and (2.26), the transmission and reflection coefficients can be rewritten in more common form of *Fresnel formulas* for TM (or *p*) polarized light:

$$r_p = \frac{n_0 \cos\theta_t - n_1 \cos\theta_i}{n_0 \cos\theta_t + n_1 \cos\theta_i}, \qquad t_p = \frac{2n_0 \cos\theta_i}{n_0 \cos\theta_t + n_1 \cos\theta_i} \tag{2.27a}$$

2.3.2 TE Polarization

The derivation of the reflection and transmission coefficients for TE-polarized waves is similar to the case of TM-polarized waves given above. Preserving the same abbreviations introduced for the TM case, according to (2.15), for TE polarization, the electric and magnetic field unit vectors (see figure 2.1b) can be written in the following form:

$$\vec{e}_i = (0,1,0),\ \vec{e}_r = (0,1,0),\ \vec{e}_t = (0,1,0),$$
$$\vec{h}_i = \frac{c}{\omega}(-k_{0z},0,k_x),\ \vec{h}_r = \frac{c}{\omega}(k_{0z},0,k_x),\ \vec{h}_t = \frac{c}{\omega}(-k_{1z},0,k_x)$$

In this case, the total electric field in the medium 1, $\boldsymbol{E}^{(0)}$, will be

$$\vec{E}^{(0)} = E_+^{(0)} \cdot e^{i \cdot k_{0z} z}(0,1,0) + E_-^{(0)} \cdot e^{-i \cdot k_{0z} z}(0,1,0) \tag{2.28a}$$

the total electric field in the medium 2, $\boldsymbol{E}^{(1)}$, will be

$$\vec{E}^{(1)} = E_{+}^{(1)} \cdot e^{i \cdot k_{1z}(z-d)}(0,1,0) \tag{2.28b}$$

The total magnetic field in the medium 1, $\boldsymbol{H}^{(0)}$, will be

$$\vec{H}^{(0)} = \frac{c}{\omega} E_{+}^{(0)} \cdot e^{i \cdot k_{0z} z}\left(-k_{0z}, 0, k_x\right) + \frac{c}{\omega} E_{-}^{(0)} \cdot e^{-i \cdot k_{0z} z}\left(k_{0z}, 0, k_x\right) \tag{2.28c}$$

and, the magnetic field in the medium 2, $\boldsymbol{H}^{(1)}$, will be

$$\vec{H}^{(1)} = \frac{c}{\omega} E_{+}^{(1)} \cdot e^{i \cdot k_{1z} z}\left(-k_{0z}, 0, k_x\right) \tag{2.28d}$$

The boundary conditions (2.19) for TE waves will take the form

$$E_y^{(0)}\Big|_{z=0} = E_y^{(1)}\Big|_{z=0} \qquad H_x^{(0)}\Big|_{z=0} = H_x^{(1)}\Big|_{z=0} \tag{2.29}$$

By substituting (2.28) into (2.29) and using some elementary math, one can find that

$$\frac{E_{-}^{(0)}}{E_{+}^{(0)}} = \frac{k_{0z} - k_{1z}}{k_{0z} + k_{1z}}, \qquad \frac{E_{+}^{(1)}}{E_{+}^{(0)}} = \frac{2k_{0z}}{k_{0z} + k_{1z}} \tag{2.30}$$

Similar to the TM case, let us define the complex reflection coefficient r_s and the complex transmission coefficient t_s:

$$r_s = \frac{E_{-}^{(0)}}{E_{+}^{(0)}}, \qquad t_s = \frac{E_{+}^{(1)}}{E_{+}^{(0)}} \tag{2.31}$$

According to (2.30) and (2.31), the transmission and reflection coefficients can be rewritten in the more common form of Fresnel formulas for TE (or s) polarized light:

$$r_s = \frac{n_0 \cos\theta_i - n_1 \cos\theta_t}{n_0 \cos\theta_i + n_1 \cos\theta_t}, \qquad t_s = \frac{2n_0 \cos\theta_i}{n_0 \cos\theta_i + n_1 \cos\theta_t} \tag{2.32}$$

Using (2.27) and (2.32), reflection and transmission coefficients can be obtained for any polarization state of an incident wave. As was mentioned above, p- and s-waves are independent from each other in both media. Hence, the arbitrarily polarized incident wave can be expanded into p- and s-components; the reflection and transmission coefficients can be found separately for both components according to (2.27) and (2.32), and the resultant waves can be recombined.

The important parameter of an electromagnetic wave is the *intensity* of the wave. For a plane electromagnetic waves in the form of $\mathbf{e} A \exp[i(\omega t - \boldsymbol{k}\cdot\boldsymbol{r})]$ intensity is commonly defined as

$$I = A \cdot A^* = |A|^2 \tag{2.33}$$

The ratio of the intensities of reflected and incident waves is usually called *reflectance* R_p (for the TM polarized wave R_s for the TE polarized wave), while the ratio of the intensities of transmitted and incident waves is usually called *transmittance*, T_p (for the TM polarized wave or T_s for the TE polarized wave):

$$R_p = r_p \cdot \bar{r}_p, \qquad R_s = r_s \cdot \bar{r}_s,$$
$$T_p = \frac{n_1 \cos\theta_t}{n_0 \cos\theta_i} t_p \cdot \bar{t}_p, \qquad T_s = \frac{n_1 \cos\theta_t}{n_0 \cos\theta_i} t_s \cdot \bar{t}_s \tag{2.34}$$

According to energy conversation law, the sum of reflectance and transmittance should be equal to unity: $R_p + T_p = 1$ and $R_s + T_s = 1$.

As an illustration, let us consider the reflection of a plane electromagnetic wave with the wavelength $\lambda = 632.8$ nm (the wavelength of the red line of a HeNe laser) from the air/glass and air/silicon boundaries. The reflection angular spectra, calculated according to (2.27), (2.32), and (2.34), are presented in figure 2.2. The dielectric permittivity of air with good accuracy is equal to unity: $\varepsilon_{air} = 1$, the dielectric permittivity of glass is assumed to be $\varepsilon_{glass} = 1.52$, and the dielectric permittivity of silicon is assumed to be $\varepsilon_{Si} = (3.882+0.019i)^2$. The complex value of the silicon refractive index at 632.8 nm wavelength indicates the presence of light absorption in silicon.

The reflectance of TE waves increases monotonically for silicon and glass substrates starting from the value of $(n-1)^2/(n+1)^2$ at the normal incidence, while for a TM polarized wave the reflectance passes the minimum value at some angle, known as *a Brewster angle*:

$$\tan\theta_B = n_1/n_0 \tag{2.35}$$

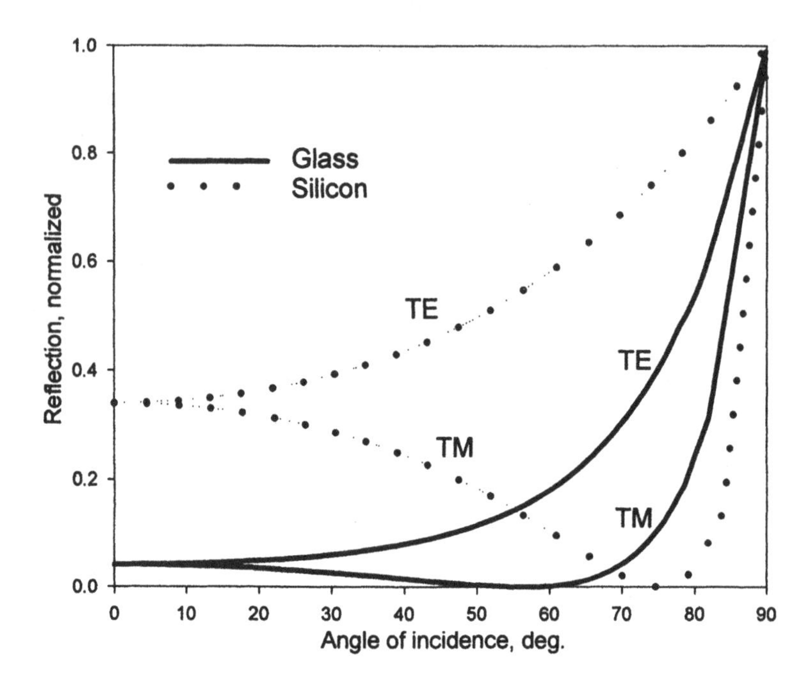

Figure 2.2. Reflection angular spectra of a single boundary for an air-glass and an air-silicon boundary at 633 nm wavelength.

A Brewster angle phenomenon serves as the basis for polarizers (i.e., devices that select a predetermined polarization of the reflected or transmitted light).

Another important single-boundary phenomenon is the *total internal reflection* (TIR). It occurs when the refractive index of the incidence medium n_0 (see figure 2.1) is greater than that of the second medium n_1, i.e., $n_0 > n_1$. In this case the angle $0° < \theta_c < 90°$ exists at which the following equation holds: $(n_0/n_1)\sin\theta_c = 1$. According to the Snell's law (2.23), for $\theta > \theta_c$ the sinus of the propagation angle of transmitted light becomes greater than unity, which appears to be absurd. In reality, no transmitted wave exists for angles of incidence greater than θ_c all energy of the incident light wave is reflected regardless of the polarization state of the electric field vector. This phenomena is known as TIR, and the critical angle is defined as

$$\theta_c = \sin^{-1}(n_1/n_2), \qquad n_0 > n_1 \qquad (2.36)$$

and is known as the TIR angle.

In figure 2.3 the reflection angular spectra of a single boundary are given for a glass-air and glass-water boundaries at 633 nm wavelength. One can see that the higher the ratio of (n_1/n_2), the smaller the value of θ_c in accordance with equation (2.36). Both TE and TM reflection coefficients rise to unity quite quickly at the angles below θ_c. It leads to a simple determination of TIR angle with relatively high accuracy. This feature is the basis of refractometers (instruments that are used to determine refractive indexes of liquids or solids).

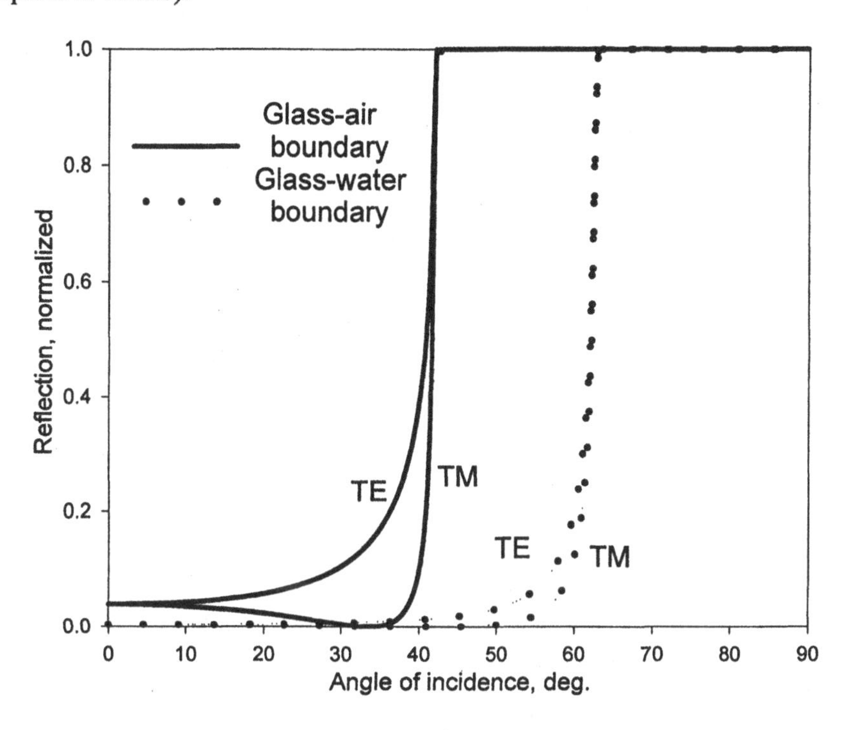

Figure 2.3. Reflection angular spectra of a single boundary for glass-air and for glass-water at 633 nm wavelength.

Total internal reflection serves as the basis for dielectric waveguides which are discussed in more detail in the following sections of this book.

2.4 The Optics of a Thin Film

Reflection and refraction of the monochromatic plane electromagnetic wave on the single boundary between two isotropic homogeneous semi-infinite media, described above, is among the simplest optical problems. However, the majority of the optical devices have both boundaries contributing to the optical performance of said device. Hence, the reflection, transmission, and absorption of light in the finite medium of different from surrounding media dielectric properties are of great importance for the optics. In this section we consider the optics of *thin film* made of an isotropic homogeneous medium. The term "*thin film*" usually includes the medium, having finite size in one direction and infinite size in the plane, perpendicular to said direction, with homogeneous dielectric properties and plane interfaces with surrounding media.

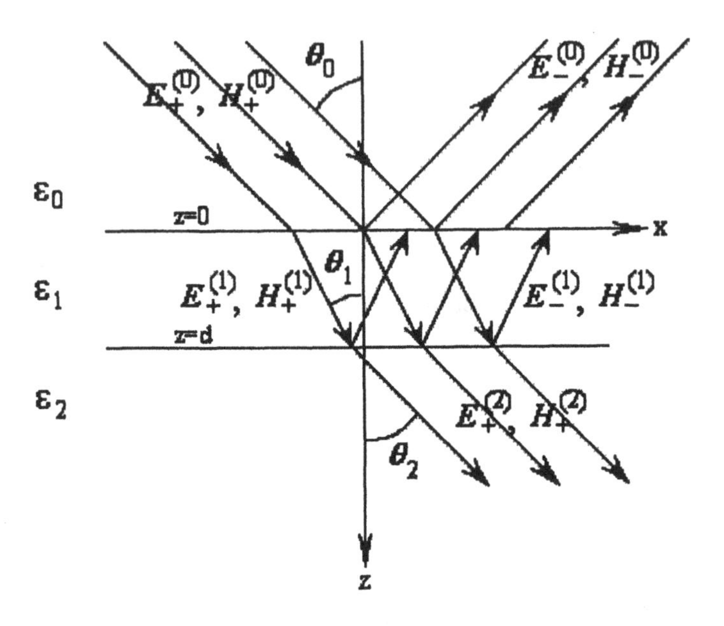

Figure 2.4. Transmission through and reflection from a single thin film.

The geometry of the problem that we are considering is given in figure 2.4. The plane monochromatic electromagnetic wave is incident from the upper medium 0, having dielectric permittivity ε_0, onto the first interface of a thin film 1, having thickness d and dielectric permittivity $\varepsilon_1 = \mathrm{Re}[\varepsilon_1] + i\cdot\mathrm{Im}\,[\varepsilon_1]$, placed on a substrate 2, having dielectric permittivity $\varepsilon_2 = \mathrm{Re}[\varepsilon_2] + i\cdot\mathrm{Im}\,[\varepsilon_2]$. Let us also assume that the incident wave is linearly polarized. As shown in figure 2.4, light will be partially reflected at the upper thin film interface, partially transmitted into the thin film, partially absorbed in the film, partially reflected into thin film at the second thin film interface, partially transmitted into substrate, and so on. To describe this process quantitatively, we will follow the approach taken above to analyze the reflection and transmission at the single boundary.

Since all media are assumed to be isotropic, using (2.20) and (2.21) at both thin film interfaces, we will obtain that all wavevectors (i.e., wavevectors of incident, reflected, transmitted into substrate 2 and propagated through thin film electromagnetic waves) must lie in a single plane – the plane of incidence. The tangential components of all mentioned wavevectors must be equal to each other. Let us define the angles between the wavevectors in the media 0,1 and 2 and the normal direction to the surface as θ_0, θ_1 and θ_2 (see figure 2.4). By using Snell law (2.23) at both thin film interfaces one can obtain

$$\frac{\sin\theta_0}{\sin\theta_1}=\frac{n_1}{n_0} \qquad \frac{\sin\theta_0}{\sin\theta_2}=\frac{n_2}{n_0} \tag{2.37}$$

This means that the direction of the transmitted wave into the substrate film does not depend on the thin film properties.

In order to find the relations of the amplitudes of transmitted and reflected waves from the amplitude of the incident wave and the part of the electromagnetic wave that will be absorbed by thin film (i.e. transformed into heat or other types of energy), we will follow the above approach and consider TM and TE polarizations of the incident light separately.

2.4.1 TM Polarization

We will use the same abbreviations as in section 2.3: the *x*- components of wavevectors in all media will be denoted as k_x . *z*-components of the wavevectors will be denoted as $k_{iz}=\sqrt{\varepsilon_i\left(\frac{\omega}{c}\right)^2-k_x^2}$, i = 0,1,2. Let us denote the incident electromagnetic wave electric field amplitude as $E_+^{(0)}$, the unit vectors collinear to electric and magnetic fidd vectors as $\mathbf{e}_+^{(0)}$ and $\boldsymbol{h}_+^{(0)}$, the amplitude of the sum of all electromagnetic waves propagating in the positive direction into the thin film by $E_+^{(1)}$, the unit vectors of electric and magnetic field of said waves by $\mathbf{e}_+^{(1)}$ and $\boldsymbol{h}_+^{(1)}$, the amplitude of the sum of all electromagnetic waves propagating in the negative direction into the thin film by $E_-^{(1)}$, the unit vectors of electric and magnetic field of said waves by $\mathbf{e}_-^{(1)}$ and $\boldsymbol{h}_-^{(1)}$, the amplitude of the electric field vector of the transmitted into substrate wave by $E_+^{(2)}$, and the unit vectors of electric and magnetic field of said wave by $\mathbf{e}_-^{(2)}$ and $\boldsymbol{h}_-^{(2)}$. In this case the electric and magnetic field unit vectors of said electromagnetic waves can be written in the following form:

$$\vec{e}_+^{(i)}=\left(1{,}0,-\frac{k_x}{k_{iz}}\right),\quad i=0,1,2; \qquad \vec{e}_-^{(i)}=\left(1{,}0,\frac{k_x}{k_{iz}}\right),\quad i=0,1;$$

$$\vec{h}_+^{(i)}=\frac{\omega}{c}\left(0,-\frac{\varepsilon_i}{k_{iz}},0\right),\ i=0,1,2; \qquad \vec{h}_-^{(i)}=\frac{\omega}{c}\left(0,\frac{\varepsilon_i}{k_{iz}},0\right),\ i=0,1.$$

The total electric field in the medium 0, $\boldsymbol{E}^{(0)}$ will be

$$\vec{E}^{(0)} = E_{+}^{(0)} \cdot e^{i\cdot k_{0z}z}\left(1,0,-\frac{k_x}{k_{0z}}\right) + E_{-}^{(0)} \cdot e^{-i\cdot k_{0z}z}\left(1,0,\frac{k_x}{k_{0z}}\right) \tag{2.38a}$$

the total electric field in the thin film, $\boldsymbol{E}^{(1)}$ will be

$$\vec{E}^{(1)} = E_{+}^{(1)} \cdot e^{i\cdot k_{1z}z}\left(1,0,-\frac{k_x}{k_{1z}}\right) + E_{-}^{(0)} \cdot e^{-i\cdot k_{1z}(z-d)}\left(1,0,\frac{k_x}{k_{1z}}\right) \tag{2.38b}$$

and the total electric field in the substrate, $\boldsymbol{E}^{(2)}$ will be

$$\vec{E}^{(2)} = E_{+}^{(2)} \cdot e^{i\cdot k_{2z}(z-d)}\left(1,0,-\frac{k_x}{k_{2z}}\right) \tag{2.38c}$$

Similarly, magnetic fields in all three medias will be

$$\vec{H}^{(0)} = \frac{\omega}{c}E_{+}^{(0)} \cdot e^{i\cdot k_{0z}z}\left(0,-\frac{\varepsilon_o}{k_{0z}},0\right) + \frac{\omega}{c}E_{-}^{(0)} \cdot e^{-i\cdot k_{0z}z}\left(0,\frac{\varepsilon_o}{k_{0z}},0\right) \tag{2.38d}$$

$$\vec{H}^{(1)} = \frac{\omega}{c}E_{+}^{(1)} \cdot e^{i\cdot k_{1z}z}\left(0,-\frac{\varepsilon_1}{k_{1z}},0\right) + \frac{\omega}{c}E_{-}^{(1)} \cdot e^{-i\cdot k_{1z}(z-d)}\left(0,\frac{\varepsilon_1}{k_{1z}},0\right) \tag{2.38e}$$

$$\vec{H}^{(2)} = \frac{\omega}{c}E_{+}^{(2)} \cdot e^{i\cdot k_{2z}(z-d)}\left(0,-\frac{\varepsilon_2}{k_{2z}},0\right) \tag{2.38f}$$

The boundary conditions (2.19) in this case will take the form

$$\begin{aligned} &\left.E_x^{(0)}\right|_{z=0} = \left.E_x^{(1)}\right|_{z=0}, \qquad \left.H_y^{(0)}\right|_{z=0} = \left.H_y^{(1)}\right|_{z=0}, \\ &\left.E_x^{(1)}\right|_{z=d} = \left.E_x^{(2)}\right|_{z=d}, \qquad \left.H_y^{(1)}\right|_{z=d} = \left.H_y^{(2)}\right|_{z=d} \end{aligned} \tag{2.39}$$

By substituting (2.38) into (2.39) and applying some elementary math, one can find that the complex reflection and transmission coefficients of the thin film for TM polarization of the incident light will be

$$r_{02}^{TM} = \frac{E_{-}^{(0)}}{E_{+}^{(0)}} = \frac{r_{01}^{TM} + r_{12}^{TM}e^{2i\cdot k_{1z}d}}{1 + r_{01}^{TM}r_{12}^{TM}e^{2i\cdot k_{1z}d}} \tag{2.40}$$

$$t_{02}^{TM} = \frac{E_+^{(2)}}{E_+^{(0)}} = \frac{t_{01}^{TM} t_{12}^{TM} e^{i \cdot k_{1z} d}}{1 + r_{01}^{TM} r_{12}^{TM} e^{2i \cdot k_{1z} d}}$$

where r_{01}^{TM} and r_{02}^{TM} are the complex reflection coefficients at the first and second interfaces of a thin film (2.26):

$$r_{i,i+1}^{TM} = \frac{\frac{\varepsilon_i}{k_{iz}} - \frac{\varepsilon_{i+1}}{k_{(i+1)z}}}{\frac{\varepsilon_i}{k_{iz}} + \frac{\varepsilon_{i+1}}{k_{(i+1)z}}}, \qquad t_{i,i+1}^{TM} = \frac{2\frac{\varepsilon_i}{k_{iz}}}{\frac{\varepsilon_i}{k_{iz}} + \frac{\varepsilon_{i+1}}{k_{(i+1)z}}}, \; i=0,1 \tag{2.41}$$

2.4.2 TE Polarization

Preserving the same abbreviations introduced for TM case, the electric and magnetic field unit vectors can be written in the following form:

$$\vec{e}_+^{(i)} = (0,1,0), \qquad i = 0,1,2; \qquad \vec{e}_-^{(i)} = (0,1,0), \qquad i = 0,1;$$
$$\vec{h}_+^{(i)} = \frac{c}{\omega}(-k_{iz}, 0, k_x), \; i = 0,1,2; \qquad \vec{h}_-^{(i)} = \frac{c}{\omega}(k_{iz}, 0, k_x), \; i = 0,1.$$

In this case, total electric fields in the media will be

$$\vec{E}^{(0)} = E_+^{(0)} \cdot e^{i \cdot k_{0z} z}(0,1,0) + E_-^{(0)} \cdot e^{-i \cdot k_{0z} z}(0,1,0) \tag{2.42a}$$
$$\vec{E}^{(1)} = E_+^{(1)} \cdot e^{i \cdot k_{1z} z}(0,1,0) + E_-^{(1)} \cdot e^{-i \cdot k_{1z}(z-d)}(0,1,0) \tag{2.42b}$$
$$\vec{E}^{(2)} = E_+^{(2)} \cdot e^{i \cdot k_{2z}(z-d)}(0,1,0) \tag{2.42c}$$

Total magnetic fields in the media will be

$$\vec{H}^{(0)} = \frac{c}{\omega} E_+^{(0)} \cdot e^{i \cdot k_{0z} z}(-k_{0z}, 0, k_x) + \frac{c}{\omega} E_-^{(0)} \cdot e^{-i \cdot k_{0z} z}(k_{0z}, 0, k_x) \tag{2.42d}$$
$$\vec{H}^{(1)} = \frac{c}{\omega} E_+^{(1)} \cdot e^{i \cdot k_{1z} z}(-k_{1z}, 0, k_x) + \frac{c}{\omega} E_-^{(1)} \cdot e^{-i \cdot k_{1z}(z-d)}(k_{1z}, 0, k_x) \tag{2.42e}$$
$$\vec{H}^{(2)} = \frac{c}{\omega} E_+^{(2)} \cdot e^{i \cdot k_{2z}(z-d)}(-k_{2z}, 0, k_x) \tag{2.42f}$$

The boundary conditions (2.19) for TE waves will take the form

$$E_y^{(0)}\Big|_{z=0} = E_y^{(1)}\Big|_{z=0}, \; H_x^{(0)}\Big|_{z=0} = H_x^{(1)}\Big|_{z=0},$$

$$E_y^{(1)}\Big|_{z=d} = E_y^{(2)}\Big|_{z=d},\ H_x^{(1)}\Big|_{z=d} = H_x^{(2)}\Big|_{z=d} \tag{2.43}$$

By substituting (2.42) into (2.43) and applying some elementary math one can find that the complex reflection and transmission coefficients of the thin film for TE polarization of the incident light will be

$$r_{02}^{TE} = \frac{E_-^{(0)}}{E_+^{(0)}} = \frac{r_{01}^{TE} + r_{12}^{TE} e^{2i\cdot k_{1z}d}}{1 + r_{01}^{TE} r_{12}^{TE} e^{2i\cdot k_{1z}d}};\ t_{02}^{TE} = \frac{E_+^{(2)}}{E_+^{(0)}} = \frac{t_{01}^{TE} t_{12}^{TE} e^{i\cdot k_{1z}d}}{1 + r_{01}^{TE} r_{12}^{TE} e^{2i\cdot k_{1z}d}} \tag{2.44}$$

where $r_{01}{}^{TE}$ and $r_{02}{}^{TE}$ are the complex reflection coefficients at the first and second interfaces of a thin film (2.32)

$$r_{i,i+1}^{TE} = \frac{k_{iz} - k_{(i+1)z}}{k_{iz} + k_{(i+1)z}},\qquad t_{i,i+1}^{TE} = \frac{2k_{iz}}{k_{iz} + k_{(i+1)z}},\ i=0,1 \tag{2.45}$$

Expressions (2.40) to (2.41) and (2.44) to (2.45) define complex reflection and transmission coefficients of a thin film for both polarizations of an incident electromagnetic wave. As for the single-boundary case discussed above, the reflection and transmission coefficients can be derived for any polarization state of the incident wave from (2.40) to (2.41) and (2.44) to (2.45). The incident wave can be expanded into *p*- and *s*- components, the reflection and transmission coefficients of these components can be found from (2.40) and (2.44), and the resultant waves can be recombined.

The reflectance R_{02}, and the transmittance T_{02} for both polarizations will be

$$R_{02} = r_{02} \cdot \bar{r}_{02},\ T_{02} = \frac{n_2 \cos\theta_2}{n_0 \cos\theta_0} t_{02} \cdot \bar{t}_{02} \tag{2.46}$$

The reflectance from the thin film is meaningful only when the medium of incidence is nonabsorbing, while the transmittance is meaningful if both the medium of incidence and the substrate are nonabsorbing. For the absorbing thin film (i.e., for $\mathrm{Im}[n_1] \neq 0$), another important optical parameter can be introduced, the *absorptance A*. It can be defined as the fraction of electromagnetic wave energy dissipated in the thin film

$$A = 1 - R_{02} - T_{02} \tag{2.47}$$

If all media are nonabsorbing, the absorptance is equal to zero and

$R_{02} + T_{02} = 1$.

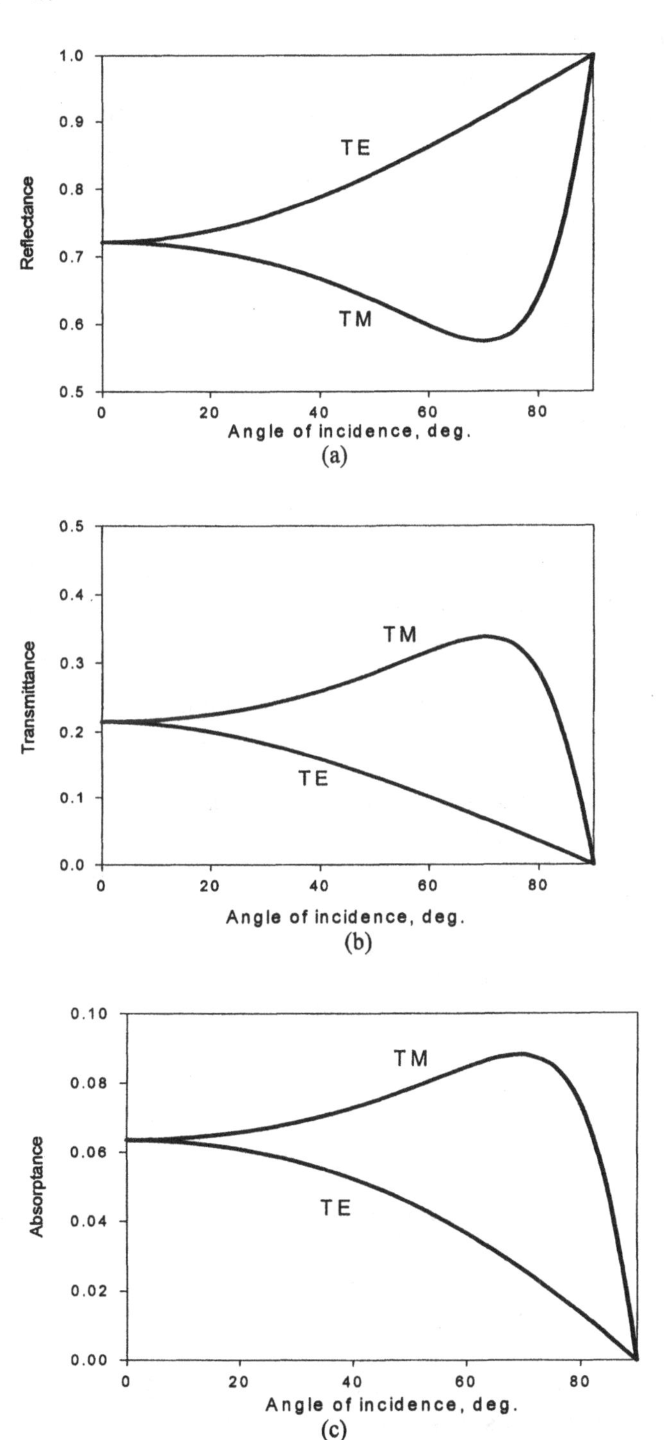

Figure 2.5. Angular dependences of the reflectance (a), transmittance (b), and absorptance (c) for a 30 nm gold film on a glass substrate at the 633 nm wavelength.

Let us consider as an illustration a 30 nm thick gold film on the glass substrate at the wavelength of an HeNe laser red line (632.8 nm). Figure 2.5 presents the numerically calculated according to (2.40) to (2.41) and (2.44) to (2.47) angular dependences of reflectance (a), transmittance (b), and absorptance (c). The reflectance, transmittance, and absorptance of a TE-polarized plane electromagnetic wave show monotonic rise (reflectance) or decrease (transmittance and absorptance), while the reflectance, transmittance, and absorptance of a TM-polarized plane electromagnetic wave passes through the minimum (reflectance) or maximum (transmittance and absorptance). It is directly related to the Brewster angle.

2.5 The Multilayer Stack

Let's now consider a more complex problem – the reflection, transmission, and absorption of a monochromatic plane electromagnetic wave in the *multilayer stack.* The term the *multilayer stack* usually refers to the number of thin films with plane boundaries are placed (by means of, for example magnetron sputtering, thermal deposition, or other common techniques) in a series. The multilayer stacks are the basis of most optical filter designs, so the problem, which will be addressed in this section is of great importance for understanding the following material in this book.

The geometry of the problem under consideration is shown in figure 2.6. The plane monochromatic electromagnetic wave is incident from the upper medium *0*, having dielectric permittivity ε_0, onto the multilayer stack, consisting of the films 1, 2..., N - 1, having thickness d_i, $i = 1,2,\ldots,N-1$ and dielectric constants $\varepsilon_i = \mathrm{Re}[\varepsilon_i] + i\cdot\mathrm{Im}\,[\varepsilon_i]$, $i = 1,2\ldots N-1$, placed on a substrate N, having dielectric permittivity $\varepsilon_N = \mathrm{Re}[\varepsilon_N] + i\cdot\mathrm{Im}\,[\varepsilon_N]$. Let us also assume that the incident wave is linearly polarized.

Since all layers in the multilayer stack are assumed to be isotropic, by using (2.20) and (2.21) at every interface, we will obtain that all wavevectors must lie in a single plane – the plane of incidence, the same as in single layer and single boundary cases. The tangential components of all wavevectors must be equal to each other. As before, we will consider TM and TE polarizations of the incident wave separately.

2.5.1 TM Polarization

We will use the same abbreviations as in sections 2.3 and 2.4: the x-components of wavevectors in all media will be denoted as k_x; z-components of the wavevectors will be denoted as $k_{iz} = \sqrt{\varepsilon_i \left(\frac{\omega}{c}\right)^2 - k_x^2}$, $i = 0, 1, 2, \ldots,$ N; the unit vectors collinear to electric and magnetic field vectors, and the amplitudes will be denoted as $\mathbf{e}_+^{(i)}$, $\boldsymbol{h}_+^{(i)}$ and $E_+^{(i)}$, $i = 0, 1, 2, \ldots,$ N for electromagnetic waves propagating in the positive direction in the ith layer, and $\mathbf{e}_-^{(i)}$, $\boldsymbol{h}_-^{(i)}$and $E_-^{(i)}$, i=0, 1, 2,…,(N-1) for electromagnetic waves propagating in the negative direction in the ith layer.

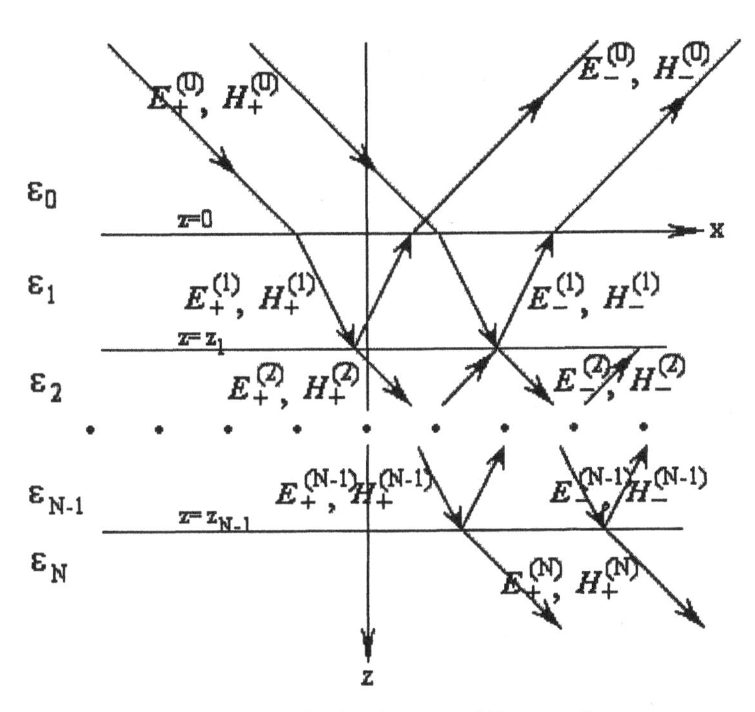

Figure 2.6. Transmission through and reflection from a multilayer stack.

In this case, the electric and magnetic field unit vectors of said electromagnetic waves can be written in the following form:

$$\vec{e}_+^{(i)} = \left(1, 0, -\frac{k_x}{k_{iz}}\right), \quad i=0,1,2..\mathrm{N}; \qquad \vec{e}_-^{(i)} = \left(1, 0, \frac{k_x}{k_{iz}}\right), \ i=0,1,2..(\mathrm{N}-1);$$

$$\vec{h}_+^{(i)} = \frac{\omega}{c}\left(0, -\frac{\varepsilon_i}{k_{iz}}, 0\right), \ i=0,1,2..\mathrm{N}; \ \vec{h}_-^{(i)} = \frac{\omega}{c}\left(0, \frac{\varepsilon_i}{k_{iz}}, 0\right), \ i=0,1,2..(\mathrm{N}-1).$$

Hence, the total electric field in each layer will be

$$\vec{E}^{(0)} = E_+^{(0)} \cdot e^{i \cdot k_{0z} z} \left(1,0,-\frac{k_x}{k_{0z}} \right) + E_-^{(0)} \cdot e^{-i \cdot k_{0z} z} \left(1,0,\frac{k_x}{k_{0z}} \right)$$

$$\vec{E}^{(1)} = E_+^{(1)} \cdot e^{i \cdot k_{1z} z} \left(1,0,-\frac{k_x}{k_{1z}} \right) + E_-^{(1)} \cdot e^{-i \cdot k_{1z}(z-z_1)} \left(1,0,\frac{k_x}{k_{1z}} \right)$$

....

$$\vec{E}^{(N)} = E_+^{(N)} \cdot e^{i \cdot k_{1z}(z-z_{N-1})} \left(1,0,-\frac{k_x}{k_{Nz}} \right)$$

where $z_i = \sum_{m=1}^{i} d_m$ - z-coordinate of ith boundary (see figure 2.6).
The total magnetic field in each layer will be

$$\vec{H}^{(0)} = \frac{\omega}{c} E_+^{(0)} \cdot e^{i \cdot k_{0z} z} \left(0,-\frac{\varepsilon_o}{k_{0z}},0 \right) + \frac{\omega}{c} E_-^{(0)} \cdot e^{-i \cdot k_{0z} z} \left(0,\frac{\varepsilon_o}{k_{0z}},0 \right)$$

$$\vec{H}^{(1)} = \frac{\omega}{c} E_+^{(1)} \cdot e^{i \cdot k_{1z} z} \left(0,-\frac{\varepsilon_1}{k_{1z}},0 \right) + \frac{\omega}{c} E_-^{(1)} \cdot e^{-i \cdot k_{1z}(z-z_1)} \left(0,\frac{\varepsilon_1}{k_{1z}},0 \right)$$

....

$$\vec{H}^{(N)} = \frac{\omega}{c} E_+^{(N)} \cdot e^{i \cdot k_{1z}(z-z_{N-1})} \left(0,-\frac{\varepsilon_N}{k_{Nz}},0 \right)$$

The boundary conditions (2.19) in this case will take the form

$$\left. E_x^{(i)} \right|_{z=z_i} = \left. E_x^{(i+1)} \right|_{z=z_i}$$

$$\text{for } i = 0,1,2..\text{N-1} \qquad (2.48)$$

$$\left. H_y^{(i)} \right|_{z=z_i} = \left. H_y^{(i+1)} \right|_{z=z_i}$$

One can note that (2.48) can be written in the matrix form:

$$A^{TM} \cdot \begin{pmatrix} E_-^{(1)} \\ E_+^{(1)} \\ E_-^{(2)} \\ E_+^{(2)} \\ \bullet \\ E_-^{(N)} \\ E_+^{(N)} \end{pmatrix} = \begin{pmatrix} -1 \\ -\dfrac{\varepsilon_0}{k_{0z}} \\ 0 \\ 0 \\ \bullet \\ 0 \\ 0 \end{pmatrix} \tag{2.49}$$

where

$$A^{TM} = \begin{bmatrix} 1 & -1 & -e^{j\cdot k_{1z}\cdot d_1} & 0 & 0 & 0 & \bullet & 0 & 0 & 0 \\ \dfrac{\varepsilon_0}{k_{0z}} & \dfrac{\varepsilon_1}{k_{1z}} & \dfrac{\varepsilon_1}{k_{1z}} e^{j\cdot k_{1z}\cdot d_1} & 0 & 0 & 0 & \bullet & 0 & 0 & 0 \\ 0 & e^{j\cdot k_{1z}\cdot d_1} & 1 & -1 & -e^{j\cdot k_{2z}\cdot d_2} & 0 & \bullet & 0 & 0 & 0 \\ 0 & \dfrac{\varepsilon_1}{k_{1z}} e^{j\cdot k_{1z}\cdot d_1} & \dfrac{\varepsilon_1}{k_{1z}} & \dfrac{\varepsilon_2}{k_{2z}} & \dfrac{\varepsilon_2}{k_{2z}} e^{j\cdot k_{2z}\cdot d_2} & 0 & \bullet & 0 & 0 & 0 \\ 0 & 0 & 0 & e^{j\cdot k_{2z}\cdot d_2} & 1 & -1 & \bullet & 0 & 0 & 0 \\ 0 & 0 & 0 & \dfrac{\varepsilon_2}{k_{2z}} e^{j\cdot k_{2z}\cdot d_2} & \dfrac{\varepsilon_2}{k_{2z}} & \dfrac{\varepsilon_3}{k_{3z}} & \bullet & 0 & 0 & 0 \\ \bullet & \bullet & \bullet & \bullet & \bullet & \bullet & \bullet & \bullet & \bullet & \bullet \\ 0 & 0 & 0 & 0 & 0 & 0 & \bullet & \dfrac{\varepsilon_{N-1}}{k_{(N-1)z}} & \dfrac{\varepsilon_{N-1}}{k_{(N-1)z}} e^{j\cdot k_{(N-1)z}d_{N-1}} & 0 \\ 0 & 0 & 0 & 0 & 0 & 0 & \bullet & e^{j\cdot k_{(N-1)z}d_{N-1}} & 1 & -1 \\ 0 & 0 & 0 & 0 & 0 & 0 & \bullet & \dfrac{\varepsilon_{N-1}}{k_{(N-1)z}} e^{j\cdot k_{(N-1)z}d_{N-1}} & \dfrac{\varepsilon_{N-1}}{k_{(N-1)z}} & \dfrac{\varepsilon_N}{k_{Nz}} \end{bmatrix}$$

The structure of A^{TM} is as follows. The $2i$th row corresponds to the electric field continuity condition at the ith interface: $E_x^{(i)}\big|_{z=z_i} = E_x^{(i+1)}\big|_{z=z_i}$. The $2i+1$th row corresponds to the magnetic field continuity condition at the ith interface: $H_y^{(i)}\big|_{z=z_i} = H_y^{(i+1)}\big|_{z=z_i}$. The $2i$th column corresponds to $E_-^{(i)}$, while the $2i+1$th column corresponds to $E_+^{(i)}$, $i = 0,\ldots,N$.

From (2.49) it is obvious that

$$\begin{pmatrix} E_-^{(1)} \\ E_+^{(1)} \\ E_-^{(2)} \\ E_+^{(2)} \\ \bullet \\ E_-^{(N)} \\ E_+^{(N)} \end{pmatrix} = \left(A^{TM}\right)^{-1} \cdot \begin{pmatrix} -1 \\ -\frac{\varepsilon_0}{k_{0z}} \\ 0 \\ 0 \\ \bullet \\ 0 \\ 0 \end{pmatrix} \tag{2.50}$$

where $(A^{TM})^{-1}$ is the inverse matrix to A^{TM} such as $(A^{TM})^{-1} A^{TM} = I$, where I is a unitary matrix: $I_{ij} = \delta_{ij}$. The operation of matrix inversion is a standard operation for most mathematical packages, so (2.50) can be solved numerically.

2.5.2 TE Polarization

Using the same abbreviations introduced for the TM case, for TE-polarization the electric and magnetic field unit vectors can be written in the following form:

$$\vec{e}_+^{(i)} = (0,1,0), \qquad i = 0,1,2,\ldots,N;$$
$$\vec{e}_-^{(i)} = (0,1,0), \qquad i = 0,1,2,\ldots,(N-1);$$
$$\vec{h}_+^{(i)} = \frac{c}{\omega}\left(-k_{iz}, 0, k_x\right), \quad i = 0,1,2,\ldots,N;$$
$$\vec{h}_-^{(i)} = \frac{c}{\omega}\left(k_{iz}, 0, k_x\right), \qquad i = 0,1,2,\ldots,(N-1).$$

Hence, the total electric field in each layer will be

$$\vec{E}^{(0)} = E_+^{(0)} \cdot e^{i \cdot k_{0z} z}(0,1,0) + E_-^{(0)} \cdot e^{-i \cdot k_{0z} z}(0,1,0)$$
$$\vec{E}^{(1)} = E_+^{(1)} \cdot e^{i \cdot k_{1z} z}(0,1,0) + E_-^{(1)} \cdot e^{-i \cdot k_{1z}(z - z_1)}(0,1,0)$$

....

$$\vec{E}^{(N)} = E_+^{(N)} \cdot e^{i \cdot k_{Nz}(z - z_{N-1})}(0,1,0)$$

The total magnetic field in each layer will be

$$\vec{H}^{(0)} = \frac{c}{\omega} E_+^{(0)} \cdot e^{i \cdot k_{0z} z}\left(-k_{0z}, 0, k_x\right) + \frac{c}{\omega} E_-^{(0)} \cdot e^{-i \cdot k_{0z} z}\left(k_{0z}, 0, k_x\right)$$

$$\vec{H}^{(1)} = \frac{c}{\omega} E_+^{(1)} \cdot e^{i \cdot k_{1z} z} \left(-k_{1z}, 0, k_x \right) + \frac{c}{\omega} E_-^{(1)} \cdot e^{-i \cdot k_{1z}(z - z_1)} \left(k_{1z}, 0, k_x \right)$$

....

$$\vec{H}^{(N)} = \frac{c}{\omega} E_+^{(N)} \cdot e^{i \cdot k_{Nz}(z - z_{N-1})} \left(-k_{Nz}, 0, k_x \right)$$

The boundary conditions (2.19) in this case will take the form

$$\left. E_y^{(i)} \right|_{z=z_i} = \left. E_y^{(i+1)} \right|_{z=z_i} \qquad \text{for } i = 0..\text{N-1} \qquad (2.51)$$

$$\left. H_x^{(i)} \right|_{z=z_i} = \left. H_x^{(i+1)} \right|_{z=z_i}$$

One can note that (2.51) can be written in the matrix form:

$$A^{TE} \cdot \begin{pmatrix} E_-^{(1)} \\ E_+^{(1)} \\ E_-^{(2)} \\ E_+^{(2)} \\ \bullet \\ E_-^{(N)} \\ E_+^{(N)} \end{pmatrix} = \begin{pmatrix} -1 \\ k_{0z} \\ 0 \\ 0 \\ \bullet \\ 0 \\ 0 \end{pmatrix} \qquad (2.52)$$

where

$$A^{TE} = \begin{bmatrix} 1 & -1 & -e^{i \cdot k_{1z} \cdot d_1} & 0 & 0 & 0 & \bullet & 0 & 0 & 0 \\ k_{0z} & k_{1z} & -k_{1z} \cdot e^{i \cdot k_{1z} \cdot d_1} & 0 & 0 & 0 & \bullet & 0 & 0 & 0 \\ 0 & e^{i \cdot k_{1z} \cdot d_1} & 1 & -1 & -e^{i \cdot k_{2z} \cdot d_2} & 0 & \bullet & 0 & 0 & 0 \\ 0 & -k_{1z} \cdot e^{i \cdot k_{1z} \cdot d_1} & k_{2z} & k_{2z} & -k_{2z} \cdot e^{i \cdot k_{2z} \cdot d_2} & 0 & \bullet & 0 & 0 & 0 \\ 0 & 0 & 0 & e^{i \cdot k_{2z} \cdot d_2} & 1 & -1 & \bullet & 0 & 0 & 0 \\ 0 & 0 & 0 & -k_{2z} \cdot e^{i \cdot k_{2z} \cdot d_2} & k_{2z} & k_{3z} & \bullet & 0 & 0 & 0 \\ \bullet & \bullet & \bullet & \bullet & \bullet & \bullet & \bullet & \bullet & \bullet & \bullet \\ 0 & 0 & 0 & 0 & 0 & 0 & \bullet & k_{(N-1)z} & -k_{(N-1)z} \cdot e^{i \cdot k_{(N-1)z} d_{N-1}} & 0 \\ 0 & 0 & 0 & 0 & 0 & 0 & \bullet & e^{i \cdot k_{(N-1)z} d_{N-1}} & 1 & -1 \\ 0 & 0 & 0 & 0 & 0 & 0 & \bullet & -k_{(N-1)z} \cdot e^{i \cdot k_{(N-1)z} d_{N-1}} & k_{(N-1)z} & k_{Nz} \end{bmatrix}$$

The structure of A^{TE} is as follows. The 2*i*th row corresponds to the electric field continuity condition at the *i*th interface: $E_y^{(i)}\big|_{z=z_i} = E_y^{(i+1)}\big|_{z=z_i}$. The 2*i*+1th row corresponds to magnetic field continuity condition at the ith interface: $H_x^{(i)}\big|_{z=z_i} = H_x^{(i+1)}\big|_{z=z_i}$. The 2*i*th column corresponds to $E_-^{(i)}$, while the 2*i*+1th column corresponds to $E_+^{(i)}$, *i*=0,…,N. From (2.53) it is obvious that

$$\begin{pmatrix} E_-^{(1)} \\ E_+^{(1)} \\ E_-^{(2)} \\ E_+^{(2)} \\ \bullet \\ E_-^{(N)} \\ E_+^{(N)} \end{pmatrix} = \left(A^{TE}\right)^{-1} \cdot \begin{pmatrix} -1 \\ k_{0z} \\ 0 \\ 0 \\ \bullet \\ 0 \\ 0 \end{pmatrix} \tag{2.53}$$

where $(A^{TE})^{-1}$ is the inverse matrix to A^{TE} such as $(A^{TE})^{-1} A^{TE} = I$, where I is a unitary matrix: $I_{ij} = \delta_{ij}$.

The reflectance from the multilayer stack R, the transmittance through the multilayer stack T, and the absorptance in the multilayer stack for both polarizations can be found from the solutions of (2.50) and (2.53), respectively:

$$R = r \cdot \bar{r} = \left|E_-^{(0)}\right|^2, \qquad T = \frac{n_N \cos\theta_N}{n_0 \cos\theta_0} t \cdot \bar{t} = \frac{n_N \cos\theta_N}{n_0 \cos\theta_0} \left|E_+^{(N)}\right|^2,$$
$$A = 1 - R - T$$

where θ_N can be found from the Snell law:

$$\sin\theta_N = \frac{n_0}{n_N} \sin\theta_0$$

By solving (2.50) and (2.53), the electric and magnetic fields in each layer can be found for the TM and TE polarizations of the incident electromagnetic wave, which is important for many optical problems. In this book, however, we focus mostly on reflection and transmission amplitudes, i.e. on $t_{TM} = E_+^{(N)}$ and $r_{TM} = E_-^{(0)}$. It turns out that $E_+^{(N)}$ and $E_-^{(0)}$ can be found alone using a much less time-consuming algorithm known as the *matrix method.* This method is beyond the scope of this book. The reader can find a thorough description of the matrix method in [18].

2.5.3 Quarter-Wave (λ/4) Stack

As an example of multilayer stack, let us consider the important case of a quarter-wave (λ/4) stack. The multilayer structure consists of pairs of alternative quarter-wave layers ($n_1 d_1 = n_2 d_2 = \lambda/4$) with refractive indices n_1 and n_2, respectively. The refractive indices of the incident medium and the substrate will be n_0 =1 (air) and n_N = 1.5 (glass), respectively. In our example, we will consider the multilayer stack to be made of alternating layers of SiO_2/(TiO_2:SiO_2) produced by, for example, the sol-gel method. We will also assume that the layers are quarter wave at the wavelength of 600 nm.

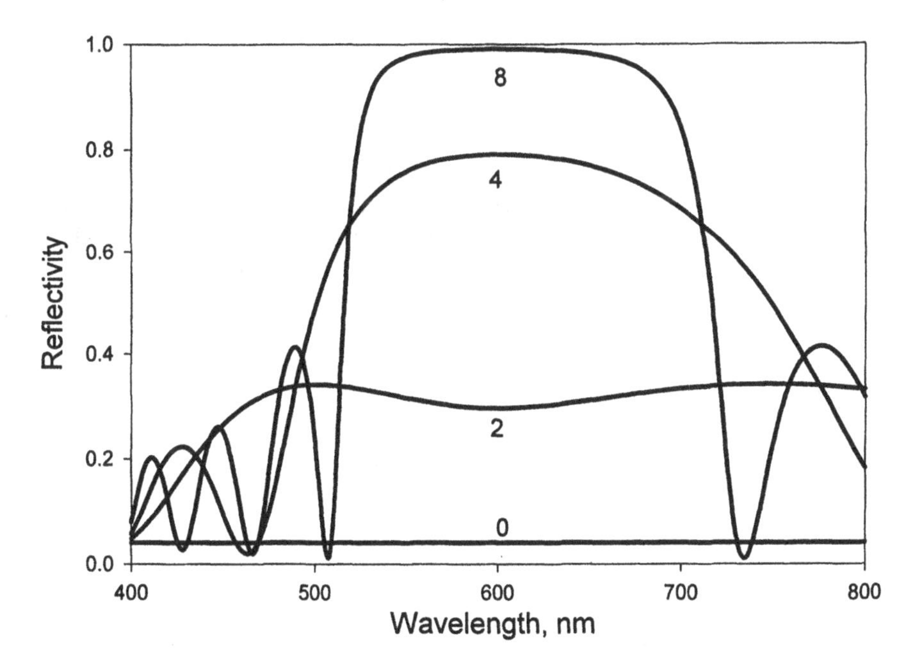

Figure 2.7. Reflectivity spectra of quarter-wave stack for a different number of SiO_2/(TiO_2:SiO_2) pairs.

The reflectivity spectra of a quarter-wave stack for various SiO_2/(TiO_2:SiO_2) pairs are given in figure 2.7. The reflectance at 600 nm approaches unity with the increase of the number of pairs, *M*. It turns out (for the derivation see, for example, [18]) that reflectivity at the quarter-wave wavelength is equal to

$$R = \tanh^2 N\chi, \qquad \chi = \ln(n_2/n_1)+(1/2M)\ln(n_N/n_0) \qquad (2.54)$$

The quarter-wave stack serves as the basis for dielectric mirrors and various filters and will be analyzed in more detail in Chapter 3 of this book.

2.6 The Optical Waveguide

In general, the term *waveguide* means the structure that supports confined electromagnetic propagation (or waveguide modes). It is known that a beam with finite transverse dimension will diverge as it propagates in a homogeneous medium. When the light wave is confined in one of the waveguide modes of the waveguide, it can propagate without divergence for hundreds of kilometers. Many types of optical waveguides exist. This section will only give a basic description of the simplest waveguide structures to provide the reader with minimal information on waveguides. That is necessary for understanding the remaining chapters of this book. Readers looking for more thorough information on waveguide theory should read more specific books on this topic (for example [20] or [5]).

2.6.1 Planar Optical Waveguide

The general requirement for guiding electromagnetic radiation is (following [19]) that there should be a flow of energy only along the guiding structure and not perpendicular to it. This means that the electromagnetic fields will be confined within the guiding structure or in the very vicinity of it.

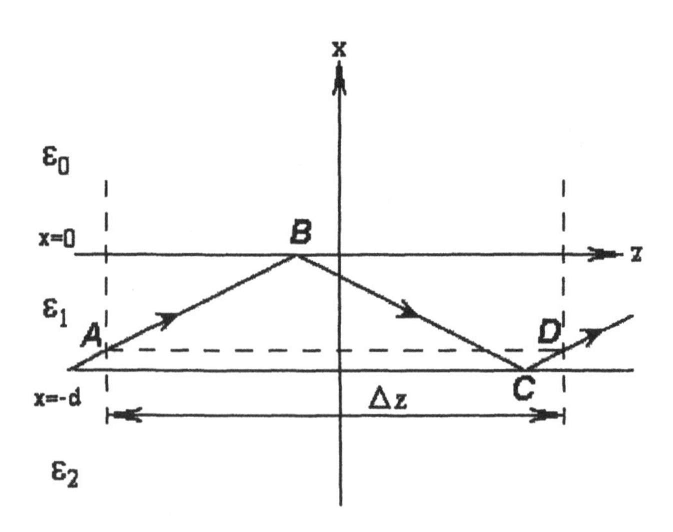

Figure 2.8. Slab optical waveguide.

Such a definition of optical waveguiding assumes a three-dimensional structure. Most waveguides used now are three-dimensional. I.e., the light is permitted to propagate only along some special direction and is localized in the plane perpendicular to that direction. However, we will start our consideration of waveguides from a two-dimensional case when the light inside the waveguide structure can propagate inside some plane rather than along some axis. Consideration of this case is much simpler than the three-dimensional case, and the understanding developed during this consideration can be easily projected into the three-dimensional case. The simplest waveguide structure of this kind is the single-layer plane dielectric waveguide, also called dielectric slab. Figure 2.8 shows a typical example of a slab waveguide. It consists of a thin dielectric layer 1 (called the *waveguiding layer* or *core*) sandwiched between two semi-infinite bounding media. Historically, the bottom semi-infinite media2 is called the *substrate*, and the top semi-infinite layer is called the *cladding* 0. The index of refraction of the waveguiding layer 1 has to be greater than those of the surrounding media. The simplest example (although not the most commercially valuable) is a symmetric slab waveguide in the form of a thin glass film immersed into air or some low-refractive index fluid.

Various methods have been developed to analyze optical waveguides. We will start our consideration of slab optical waveguides with the *ray-optics method* (*ROM*). ROM is clearly not the simplest one, and it is hard to extrapolate from it to a multilayer waveguide case, which will be considered in the next section of this book. However, ROM gives the reader, who is unfamiliar with waveguides, the simplest and most intuitive understanding of waveguiding phenomena.

The application of ROM to the analyses of the slab waveguide is possible since such a waveguide structure consists of homogeneous media with well-defined planar boundaries. As was mentioned above, light confinement in the waveguiding layer is possible only if $n_1>n_0,n_2$. It happens since the planar light waves, propagating through the waveguiding layer, experience *total internal reflections* (*TIR*) at both interfaces of the waveguiding layer 1. TIR at both interfaces of the waveguiding layer is the necessary condition for the waveguiding; although it is not sufficient to guarantee it. Let us imagine the plane wave traveling through the waveguiding layer through reflections of both its interfaces:

$$\boldsymbol{E}(x, z, t) = \boldsymbol{E}_0 \cdot \exp[i(\omega \cdot t - k_{1z} \cdot z - k_{1x} \cdot x)]$$

where $\boldsymbol{E}_0$ is a constant, k_{1x} is the transverse wave number, and k_{1z} is the parallel wave number. Let us consider that the plane wave travels the distance Δz in a time Δt in one full zigzag *ABCD* (figure 2.8). The total phase shift the plane wave will experience while traveling through *ABCD* will be the sum of the phase shifts due to propagation through *AB*, *BC*, *CD*, and TIRs at points *B* and *C*:

$$\omega \cdot \Delta t - k_{1z} \cdot \Delta z - 2 k_{1x} \cdot d + 2\varphi_{10} + 2\varphi_{12}$$

where $2\varphi_{10}$ is the phase shift on TIR at *B* and $2\varphi_{12}$ is the phase shift on TIR at *C*. Waveguide mode also can be presented in the form of $\boldsymbol{E}^w(x, z, t) = \boldsymbol{E}_0^w(x) \cdot \exp[i(\omega \cdot t - (\omega / c) \cdot n^* \cdot z)]$, where n^* is the waveguide mode *effective refractive index* and $\boldsymbol{E}_0^w(x)$ is the mode electric field distribution across the waveguide cross-section, which should be constant along *z* direction in the uniform waveguide. Such a mode will experience the phase shift of $\omega \cdot \Delta t - (\omega / c) \cdot n^* \cdot \Delta z$ while traveling from point *A* to point *D*. Hence, a totally reflecting zigzag ray will become the waveguide mode only when the sum of an extra transverse phase shift is an integer multiple of 2π:

$$-2 k_{1x} \cdot d + 2\varphi_{10} + 2\varphi_{12} = -2m\pi \tag{2.55}$$

where *m* is an integer. The phase shifts $2\varphi_{10}$ and $2\varphi_{12}$ are different for different polarizations of light and can be expressed in the following form:

$$2\varphi_{10} = \begin{cases} 2\tan^{-1}\dfrac{k_{0z}}{k_{1z}} & (TE) \\ 2\tan^{-1}\dfrac{n_1^2 k_{0z}}{n_0^2 k_{1z}} & (TM) \end{cases} \tag{2.56a}$$

$$2\varphi_{12} = \begin{cases} 2\tan^{-1}\dfrac{k_{2z}}{k_{1z}} & (TE) \\ 2\tan^{-1}\dfrac{n_1^2 k_{2z}}{n_2^2 k_{1z}} & (TM) \end{cases} \tag{2.56b}$$

The fundamental mode corresponds to the solution of (2.55) with $m = 0$, while higher-order modes correspond to $m = 1, 2, 3, \ldots$ It should be noted that negative values of *m* do not represent any physical solution. Equations (2.55) and (2.56) can be rewritten into the more common form [19] for TE modes:

$$\tan(hd) = \frac{p+q}{h(1 - {pq}/{h^2})} \tag{2.57a}$$

where $h = (\omega/c)[n_1^2-(n^*)^2]^{1/2}$; $q = (\omega/c)[(n^*)^2-n_0^2]^{1/2}$; $p = (\omega/c)[(n^*)^2-n_2^2]^{1/2}$; and for TM modes:

$$\tan(hd) = \frac{h(p'+q')}{h^2 - p'q'} \tag{2.57b}$$

where $p' = (n_1/n_2)^2 p$ and $q' = (n_1/n_0)^2 q$.

Equations (2.57a) and (2.57b) cannot be solved analytically. Graphical or numerical methods are usually used to solve these equations. Descriptions of the graphical method of solving can be found in for example [20] and [5]. This method was widely practiced about thirty years ago, but since the arrival of the computers has been mostly replaced by numerical methods. As an illustration, let us consider the practical and important case of silicon on silica waveguide. The silica substrate is assumed to have a refractive index $n_2 = 1.44$, the silicon waveguiding layer is assumed to be $n_1 = 3.5$, and the cladding is assumed to be air with $n_0 = 1$. The calculations were made around 1550 nm, so the absorption of silicon and silicon dioxide was neglected. The dispersion curves for such a waveguide calculated numerically according to equations (2.57a) and (2.57b) are given in figure 2.9.

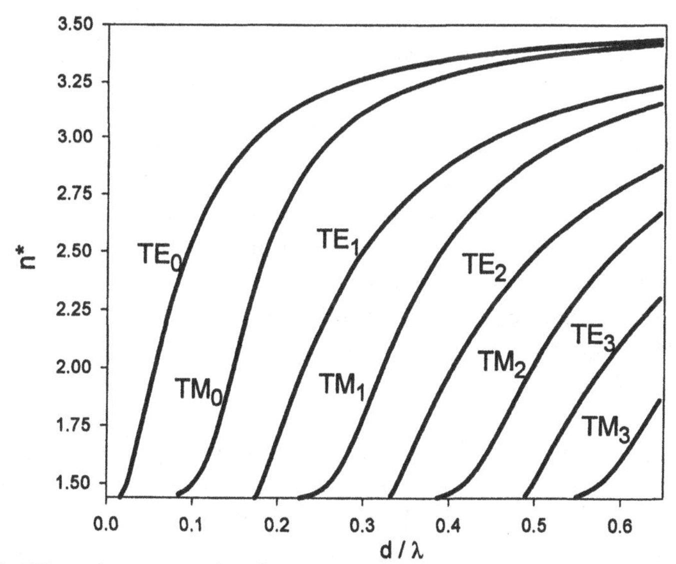

Figure 2.9. Dispersion curves for the waveguide modes of silicon waveguide on silicon dioxide substrate: n_0=1, n_1=3.5, n_2=1.44.

In general, a waveguide mode appears (i.e., the light wave became confined) above a certain value of d/λ, which is called *mode cutoff.* At or below the cutoff value, the mode extends to $x = \pm\infty$ i.e., the mode becomes unconfined and cannot be called a waveguide mode anymore. This means that for any given waveguide structure there is a certain wavelength for each mode, above which this mode vanishes, and the reverse, for any given wavelength there is waveguiding layer thickness, below which this mode does not exist. For the above example, the TE_0 mode, also called the *TE fundamental mode*, for 1550 nm wavelength, exists starting from the silicon layer thickness ~ 25 nm, while the TM_0 mode, also called the *TM fundamental mode*, for 1550 nm wavelength, exists starting from the silicon layer thickness ~ 80 nm (see figure 2.9). According to mode conditions (2.57a) and (2.57b), the cutoff values of d/λ are given by

$$\left(\frac{d}{\lambda}\right)_{TE} = \frac{1}{2\pi\sqrt{n_1^2 - n_2^2}}\left[m\pi + \tan^{-1}\left(\frac{n_2^2 - n_0^2}{n_1^2 - n_2^2}\right)^{1/2}\right] \tag{2.58a}$$

$$\left(\frac{d}{\lambda}\right)_{TM} = \frac{1}{2\pi\sqrt{n_1^2 - n_2^2}}\left[m\pi + \tan^{-1}\frac{n_1^2}{n_0^2}\left(\frac{n_2^2 - n_0^2}{n_1^2 - n_2^2}\right)^{1/2}\right] \tag{2.58b}$$

where m = 0, 1, 2, 3,... corresponds to the waveguide mode numbers (i.e., TE_m or TM_m). As follows from (2.58), the number of modes supported by the waveguide structure with a given ratio d/λ depends on the refractive index contrasts between the waveguiding layer and surrounding layers. In general, the higher the contrasts, the more waveguide modes exist for the given value of d/λ or, reciprocally, less thickness of waveguiding layer is required for the given wavelength to support a given number of waveguide modes. To illustrate this, let us compare the following examples: for a typical communication fiber constructed by creating ~ 9 microns core of doped glass higher-refractive index region inside lower-refractive index glass cladding, the refractive index contrast is usually within 0.1%. This leads to a fundamental mode cutoff value of d at 1550 nm of about 5 to 7 microns, while in the above case of silicon on silica waveguide 230% contrast at Si/SiO_2 and 350% contrast on Si/air interface led to the fundamental mode cut-off value of ~20 nm.

For any given waveguide structure and given wavelength, the waveguide modes that correspond to this structure have a distinct electric field distribution. In figure 2.10, the TE modes electric field distributions are given for the three lowest-order modes for a 700 nm thick silicon layer (a) and the TM modes electric field distributions for the three lowest-order modes (b).

Figure 2.10 shows that the confinement of the waveguide modes is decreased with the mode number and the confinement of the TM mode is generally weaker than the confinement of the TE mode for the same mode number. According to figure 2.10, the fraction of mode power of each of the waveguide mode is higher in silica than in air. This is the general property of the waveguide modes; the mode field localization is generally higher in higher refractive index layers than in lower refractive index layers. Consequently, the waveguide mode electric (and magnetic) field distribution is asymmetric in the asymmetric waveguide structure and symmetric in symmetric waveguide structure.

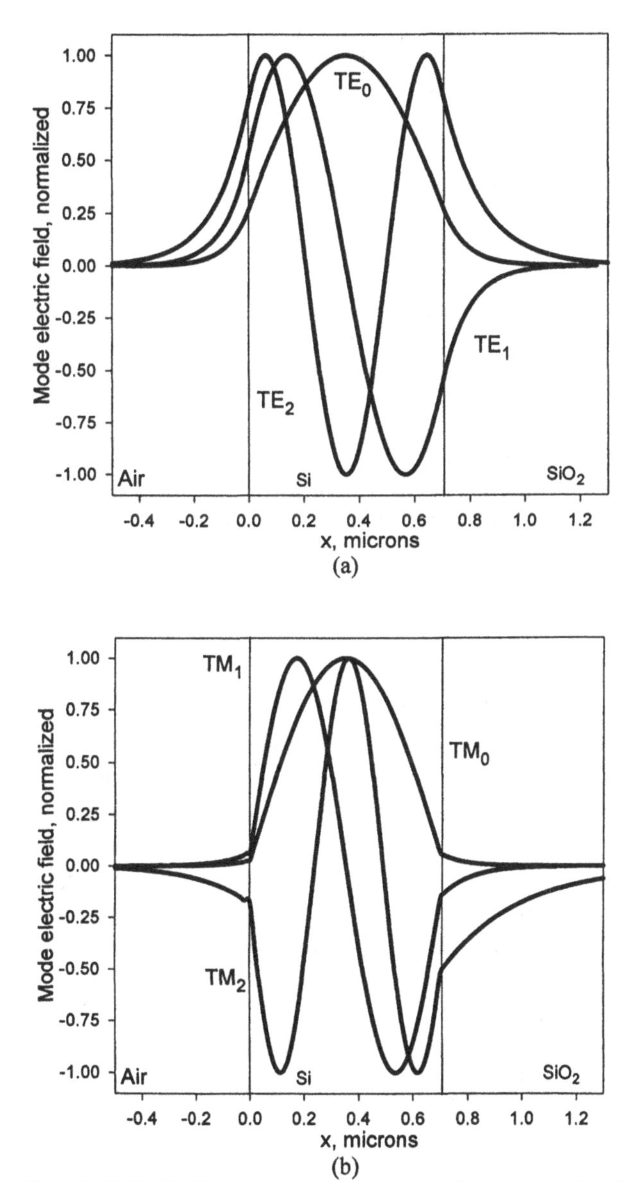

Figure 2.10. Electric field distribution across the waveguide cross-section for TE (a) and TM (b) modes.

Another waveguide feature illustrated by figure 2.10 is the different number of the a mode electric field peaks for different mode numbers. For fundamental modes (both TE and TM), the electric field distributions have the near bell-shape with just one peak; for first order modes there are two peaks (one positive and one negative), while for the second order modes there are three peaks (two positive and one negative for TE and one positive and two negative for TM polarization). This is the general property of the waveguide modes; the number of points across the waveguide cross-section, where mode field is equal to 0 is equal to the number of mode m: i.e., for 0-order mode such points are absent, for first there is one point, for second – 2, for third –3 and so on.

An important property of waveguide modes is their orthogonality:

$$C_{lm}^{TE} \int \boldsymbol{E}_l(x) \cdot \boldsymbol{E}_m^*(x)\, \mathrm{d}\, x = \delta_{\mathrm{lm}} \qquad (TE)$$

$$C_{lm}^{TM} \int \boldsymbol{H}_l(x) \cdot \boldsymbol{H}_m^*(x)\, \mathrm{d}\, x = \delta_{\mathrm{lm}} \qquad (TM) \qquad (2.59)$$

where C_{lm}^{TE} and C_{lm}^{TM} are some constants defined by the normalization of modes and δ_{lm} is the Kronecker delta. In addition, for the waveguide structure consisting of the isotropic materials, all the TE and TM modes are mutually orthogonal:

$$C_{lm} \int (\boldsymbol{E}_l(x) \times \boldsymbol{H}_m^*(x)) \cdot \mathbf{z}\, \mathrm{d}\, x = \delta_{\mathrm{lm}}$$

where C_{lm} is some constant, defined by normalization, and $\mathbf{z}$ is the unit vector in the z-axis direction. It means that for any given waveguide structure and any given wavelength, there is one and only one set of waveguide modes, supported by this structure, and any light wave, propagating through the waveguide structure, which can be expanded into the set of independent mode with the unique relative power weights. Another consequence of waveguide mode orthogonality is that each mode supported by the waveguide structure can be considered independently of the presence of other modes.

An important parameter of optical waveguides is the *numerical aperture* (*NA*). NA is defined as the largest angle of incident ray that has TIR in the waveguiding layer (or the core). For a symmetrical waveguide i.e., when $n_0 = n_2$

$$NA = \sin \theta_{\max} \approx [(n_1)^2 - (n_0)^2]^{1/2} \qquad (2.60)$$

Rays launched outside the θ_{max}, will not excite the waveguide mode. If we define the relative refractive index difference between core and cladding in a symmetrical waveguide as $\Delta = (n_1 - n_0)/n_1$, then the *NA* is related to Δ as $NA = \sin\theta_{max} \approx n_1(2\Delta)^{1/2}$. The maximum angle for the propagating light within the core (or waveguiding layer) is $\sin\varphi_{max} \approx \theta_{max}/n_1$. As the index contrast for the different waveguide types varies from 1% in communication fibers to ~ 60% in silicon on silica waveguides, *NA* and θ_{max} also vary in wide ranges from 12 deg. for single-mode optical fibers to whole $\pi/2$ for high refractive index waveguides.

2.6.2 Multilayer Optical Waveguide

Finding the effective refractive indices of a multilayers using the ROM approach given in previous section can be quite time consuming, especially if the number of layers of the multilayer is high, as in quantum well lasers or as in filter structures; which are discussed later in the text.

To construct an effective and time-efficient algorithm of a multilayer effective refractive index calculation, the matrix method is usually used (see, for example, [21]). Here I present some modifications of such a method which is not the fastest from the calculation viewpoint; although it does have significant advantages in simplicity of understanding and can be easily adjusted for calculations of anisotropic and gyrotropic structures.

This method can be understood by considering the electrical and magnetic fields of the electromagnetic waves in the multilayer structure (as was done in section 2.5 above, when reflection and transmission coefficients were calculated). The difference is that in waveguide mode calculations no incident from the side of the waveguide waves exists i.e., $\boldsymbol{E}_+^{(0)} = 0$, $\boldsymbol{H}_+^{(0)} = 0$ in figure 2.6 (section 2.5 of this book).

The electric and magnetic fields together with the coordinate system of multilayer waveguide are shown in figure 2.11. As in the reflection and transmission calculations case (see section 2.5), it is worthwhile considering the TE and TM mode cases separately for isotropic structure.

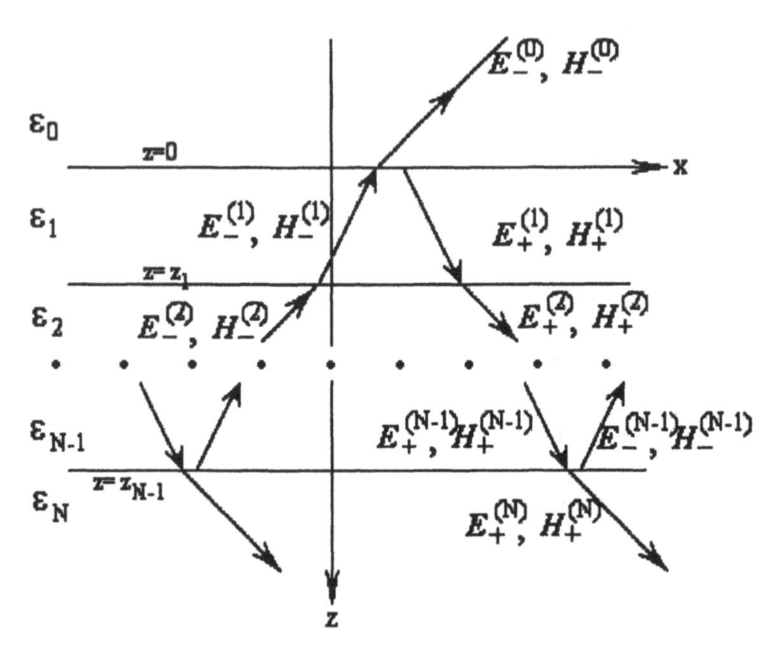

Figure 2.11. Electric and magnetic fields in the multilayer waveguide used to calculate waveguide mode characteristics.

2.6.2.1 TM Modes

The characteristic matrix of the multilayer waveguide of Figure 2.11 will have this form:

$$
A^{TM}=\begin{bmatrix}
1 & -1 & -e^{j\cdot k_{1z}\cdot d_1} & 0 & 0 & 0 & \bullet & 0 & 0 & 0 \\
\frac{\varepsilon_0}{k_{0z}} & \frac{\varepsilon_1}{k_{1z}} & \frac{\varepsilon_1}{k_{1z}}e^{j\cdot k_{1z}\cdot d_1} & 0 & 0 & 0 & \bullet & 0 & 0 & 0 \\
0 & e^{j\cdot k_{1z}\cdot d_1} & 1 & -1 & -e^{j\cdot k_{2z}\cdot d_2} & 0 & \bullet & 0 & 0 & 0 \\
0 & \frac{\varepsilon_1}{k_{1z}}e^{j\cdot k_{1z}\cdot d_1} & \frac{\varepsilon_1}{k_{1z}} & \frac{\varepsilon_2}{k_{2z}} & \frac{\varepsilon_2}{k_{2z}}e^{j\cdot k_{2z}\cdot d_2} & 0 & \bullet & 0 & 0 & 0 \\
0 & 0 & 0 & e^{j\cdot k_{2z}\cdot d_2} & 1 & -1 & \bullet & 0 & 0 & 0 \\
0 & 0 & 0 & \frac{\varepsilon_2}{k_{2z}}e^{j\cdot k_{2z}\cdot d_2} & \frac{\varepsilon_2}{k_{2z}} & \frac{\varepsilon_3}{k_{3z}} & \bullet & 0 & 0 & 0 \\
\bullet & \bullet & \bullet & \bullet & \bullet & \bullet & \bullet & \bullet & \bullet & \bullet \\
0 & 0 & 0 & 0 & 0 & 0 & \bullet & \frac{\varepsilon_{N-1}}{k_{(N-1)z}} & \frac{\varepsilon_{N-1}}{k_{(N-1)z}}e^{i\cdot k_{(N-1)z}d_{N-1}} & 0 \\
0 & 0 & 0 & 0 & 0 & 0 & \bullet & e^{i\cdot k_{(N-1)z}d_{N-1}} & 1 & -1 \\
0 & 0 & 0 & 0 & 0 & 0 & \bullet & \frac{\varepsilon_{N-1}}{k_{(N-1)z}}e^{i\cdot k_{(N-1)z}d_{N-1}} & \frac{\varepsilon_{N-1}}{k_{(N-1)z}} & \frac{\varepsilon_N}{k_{Nz}}
\end{bmatrix}
$$

The structure of A^{TM} is the same as in section 2.5.1: the 2*i*th row corresponds to the electric field continuity condition at the *i*th interface: $E_x^{(i)}\Big|_{z=z_i} = E_x^{(i+1)}\Big|_{z=z_i}$, while 2*i*+1th row corresponds to the magnetic field continuity condition at the *i*th interface: $H_y^{(i)}\Big|_{z=z_i} = H_y^{(i+1)}\Big|_{z=z_i}$, the $2i^{th}$ column corresponds to $E_-^{(i)}$, while 2*i*+1th column corresponds to $E_+^{(i)}$, $i = 0,...,$N. In this case $k_{iz} = \frac{\omega}{c}\sqrt{n_i^2 - (n^*)^2}$, where n^* is the same as before the effective refractive index of the waveguide mode.

The electromagnetic fields can have nonzero values in absence of an incident wave (it was assumed that $\boldsymbol{E}_+^{(0)} = 0$, $\boldsymbol{H}_+^{(0)} = 0$) only if det $|A| = 0$. Hence, the fundamental equation to be numerically solved for an arbitrary multilayer waveguide will be

$$\det |A^{TM}| = 0 \qquad (2.61a)$$

It should be noted that equation (2.61a) will also give the solutions of other types of modes, like surface plasmon modes (if metal layers are present in multilayer stack), leaky waveguide modes (see section 2.8 of this book), etc. Hence, the limiting conditions for the roots of (2.61a) should be set as

$$\max \{\mathrm{Re}[n_i]\} > \mathrm{Re}[n^*] > \min \{\mathrm{Re}[n_i]\} \qquad (2.61b)$$

Solutions of (2.61a) that are within the limits of (2.61b) will provide the set of TM waveguide modes, supported by multilayer waveguide structure.

2.6.2.2TE Modes

The TE modes case is similar to the TM case described above. The only difference is the structure of the characteristic matrix of the multilayer:

$$A^{TE}=\begin{bmatrix} 1 & -1 & -e^{ik_{1z}d_1} & 0 & 0 & 0 & \bullet & 0 & 0 & 0 \\ k_{0z} & k_{1z} & -k_{1z}\cdot e^{ik_{1z}d_1} & 0 & 0 & 0 & \bullet & 0 & 0 & 0 \\ 0 & e^{ik_{1z}d_1} & 1 & -1 & -e^{ik_{2z}d_2} & 0 & \bullet & 0 & 0 & 0 \\ 0 & -k_{1z}\cdot e^{ik_{1z}d_1} & k_{2z} & k_{2z} & -k_{2z}\cdot e^{ik_{2z}d_2} & 0 & \bullet & 0 & 0 & 0 \\ 0 & 0 & 0 & e^{ik_{2z}d_2} & 1 & -1 & \bullet & 0 & 0 & 0 \\ 0 & 0 & 0 & -k_{2z}\cdot e^{ik_{2z}d_2} & k_{2z} & k_{3z} & \bullet & 0 & 0 & 0 \\ \bullet & \bullet & \bullet & \bullet & \bullet & \bullet & \bullet & \bullet & \bullet & \bullet \\ 0 & 0 & 0 & 0 & 0 & 0 & \bullet & k_{(N-1)z} & -k_{(N-1)z}\cdot e^{ik_{(N-1)z}d_{N-1}} & 0 \\ 0 & 0 & 0 & 0 & 0 & 0 & \bullet & e^{ik_{(N-1)z}d_{N-1}} & 1 & -1 \\ 0 & 0 & 0 & 0 & 0 & 0 & \bullet & -k_{(N-1)z}\cdot e^{ik_{(N-1)z}d_{N-1}} & k_{(N-1)z} & k_{Nz} \end{bmatrix}$$

The structure of A^{TE} is the same as in section 2.5.2: the 2*i*th row corresponds to the electric field continuity condition at the *i*th interface: $E_y^{(i)}\big|_{z=z_i} = E_y^{(i+1)}\big|_{z=z_i}$, while the 2*i*+1th row corresponds to the magnetic field continuity condition at the *i*th interface: $H_x^{(i)}\big|_{z=z_i} = H_x^{(i+1)}\big|_{z=z_i}$, the 2*i*th column corresponds to $E_-^{(i)}$, while the 2*i*+1th column corresponds to $E_+^{(i)}$, i = 0,...,N. $k_{iz} = \frac{\omega}{c}\sqrt{n_i^2-(n^*)^2}$, where n^* is the same as before the effective refractive index of the waveguide mode. Equation (1.61a) will look like

$$\det |A^{TE}| = 0 \qquad (2.62)$$

while conditions (2.61b) hold for both TE and TM polarizations.

As an example of the implementation of such a method, let us consider the gradient-index waveguide. Gradient-index waveguides usually presume such waveguides, in which the refractive index profile changes gradually across its cross-section rather than by steps. Planar gradient index waveguides can be fabricated by the ion exchange for example. During the ion-exchange process, the ions are inserted into glass by an electrochemical process (for example, Ag^+, Na^+, K^+, or others) causing an increase of the refractive index of glass in the slide regions, where the concentration of ions is higher. Although special analytical methods have been developed for such waveguides (see, for example [22]), the method presented above can also be used here.

The refractive index distribution for a typical ion-exchanged waveguide at 632.8 nm wavelength is shown in figure 2.12a. The refractive index distribution was assumed to be Δn (z) = Δn erfc (z/d), with $\Delta n = 0.0082$, $d = 2(D_e t)^{1/2}$, $D_e = 0.03$ μm²/min; t (ion exchange time) for figure 2.12 was assumed to be 12 hours, as for Fisher premium microscope slides in 400°C potassium nitrate melt (the data were taken from [23]).

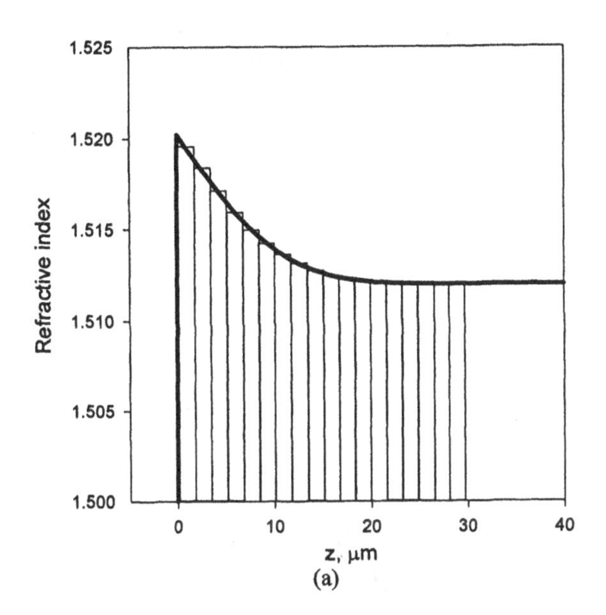

(a)

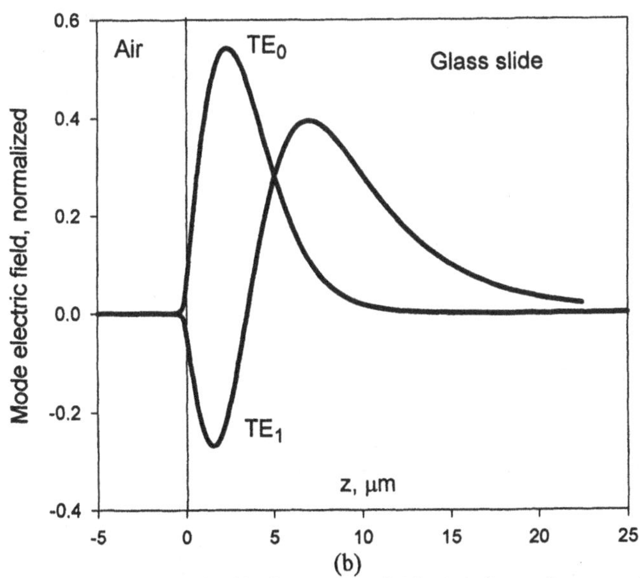

(b)

Figure 2.12. Refractive index distribution across the ion-exchanged waveguide cross-section (a); and TE waveguide mode electric field distribution across the same waveguide structure (b).

To analyze this structure, one can divide it into a number of thin layers. The refractive index change in each layer can be neglected with good accuracy (as shown in figure 2.12a) and it can be calculated using (2.61) and (2.62). Electric field distributions across the gradient index waveguide structure of Figure 2.12a are given in figure 2.12b for TE waveguide modes. For such a waveguide, most of the mode energy is localized in the microscope slide, and only a small part of it is localized in the air. In figure 2.13 the dependence of the waveguide mode effective refractive indices on the ion exchange time is shown. According to figure 2.13, for the ion-exchange process parameters given above, the ion exchange time should be roughly within 6 and 10 hours in order to get a single mode gradient index waveguide.

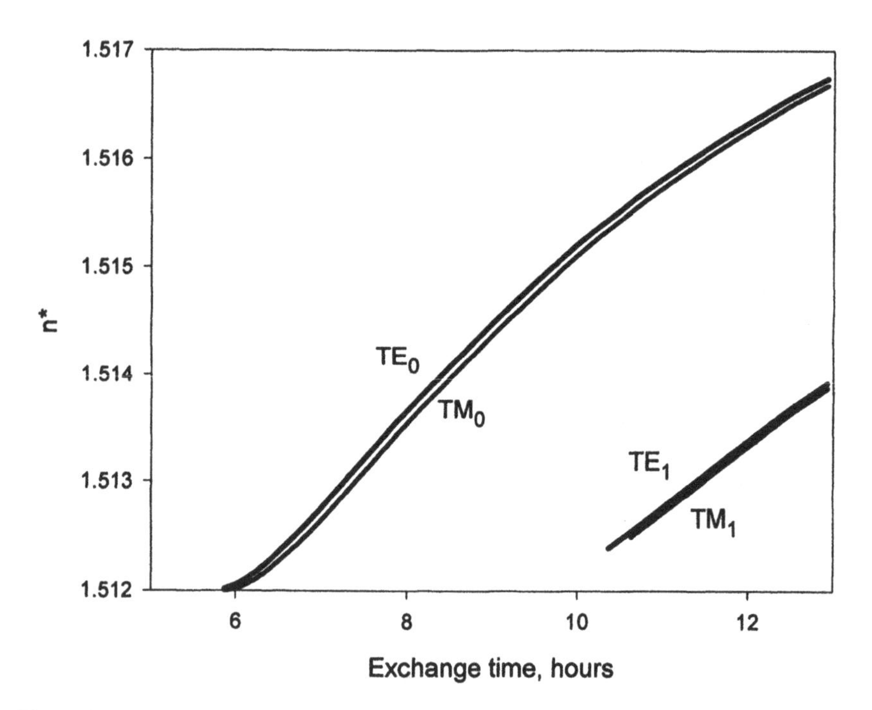

Figure 2.13. Dispersion curves for the ion-exchanged waveguide modes.

2.7 Optical Fiber

Section 2.6 discussed the planar optical waveguides (i.e., two-dimensional waveguides). However, in real life (and in the following chapters of this book) most of the waveguides used are three-dimensional. Examples of three-dimensional waveguide structures include rectangular waveguides, ridge waveguides, and cylindrical waveguides. In this book just the cylindrical waveguides, the most common example of which is optical fiber, are analyzed. For more information on this and other types of three-dimensional waveguides, readers might refer to, for example, [5].

Here the analysis focuses on step-index optical fiber with a circular cross-section, since it is the simplest case of cylindrical waveguides.
Step-index optical fiber (see figure 2.14) consists of a higher refractive index core with a refractive index n_1 and a radius a surrounded by lower refractive index cladding with a refractive index n_0 and a diameter that can be considered to be infinite with good accuracy.

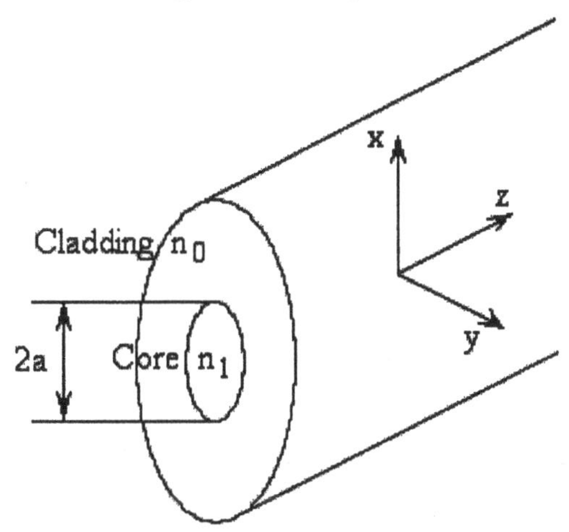

Figure 2.14. Structure of step-index optical fiber.

Due to the cylindrical symmetry of the fiber the electromagnetic fields in optical fiber in cylindrical coordinates can be written in the following form:

$$\tilde{\vec{E}} = \vec{E}(r,\theta) \cdot e^{i(\omega \cdot t - \beta \cdot z)} \qquad \tilde{\vec{H}} = \vec{H}(r,\theta) \cdot e^{i(\omega \cdot t - \beta \cdot z)} \tag{2.63}$$

where b is the propagation constant of the waveguide mode. By substitution (2.63) into the Maxwell equation one can obtain the system of wave equations [24]:

$$\begin{cases} \dfrac{\partial^2 E_z}{\partial r^2} + \dfrac{1}{r}\dfrac{\partial E_z}{\partial r} + \dfrac{1}{r^2}\dfrac{\partial^2 E_z}{\partial \theta^2} + \left[k^2 n(r,\theta)^2 - \beta^2\right] E_z = 0 \\ \dfrac{\partial^2 H_z}{\partial r^2} + \dfrac{1}{r}\dfrac{\partial H_z}{\partial r} + \dfrac{1}{r^2}\dfrac{\partial^2 H_z}{\partial \theta^2} + \left[k^2 n(r,\theta)^2 - \beta^2\right] H_z = 0 \end{cases} \tag{2.64}$$

Since we are considering axially symmetric optical fibers, the refractive index is independent of θ. Moreover, we consider here only the simplest case of axially symmetric fibers, where the refractive index is constant over the core (n_1) and over the cladding (n_0). We will consider TE and TE modes separately.

2.7.1 TE Modes

Since for TE modes $E_z = 0$ in (2.64), the system of wave equations can be written in the following form:

$$\frac{\partial^2 H_z}{\partial r^2} + \frac{1}{r}\frac{\partial H_z}{\partial r} + \left[k^2 n(r)^2 - \beta^2\right] H_z = 0$$

$$\begin{cases} E_r = 0 \\ E_\theta = \dfrac{i \cdot \omega}{k^2 n(r)^2 - \beta^2} \dfrac{\partial H_z}{\partial r} \\ H_r = -\dfrac{i \cdot \beta}{k^2 n(r)^2 - \beta^2} \dfrac{\partial H_z}{\partial r} \\ H_\theta = 0 \end{cases} \tag{2.65}$$

The solution of (2.65) is (for a detailed derivation see, for example, [5])

$$H_z = \begin{cases} A \cdot J_0(\kappa \cdot r) & (0 \le r \le a) \\ B \cdot K_0(\sigma \cdot r) & (r > a) \end{cases} \tag{2.66}$$

where $\kappa = \sqrt{k^2 n_1^2 - \beta^2}$, $\sigma = \sqrt{\beta^2 - k^2 n_0^2}$, J_0 is the 0-order Bessel function, K_0 is the modified 0-order Bessel function of second kind, and A and B are constants, connected with each other by the boundary conditions of continuity of H_z and E_θ at $r = a$.

The dispersion relation for TE modes can be derived from boundary conditions:

$$\frac{J_1(u)}{u \cdot J_0(u)} = -\frac{K_1(w)}{w \cdot K_0(w)} \tag{2.67}$$

where $u = a \cdot \kappa = a\sqrt{k^2 n_1^2 - \beta^2}$, $w = a \cdot \sigma = a \cdot \sqrt{\beta^2 - k^2 n_0^2}$. Equation (2.67) does not have an analytical solution. However, it can be solved relatively easily numerically with available mathematical packets.

The electromagnetic fields in the TE mode are

$$\begin{cases} E_r = 0 \\ E_\theta = -i \cdot \omega \cdot \dfrac{a}{u} \cdot A \cdot J_1\left(\dfrac{u}{a} r\right) \\ E_z = 0 \\ H_r = i \cdot \beta \cdot \dfrac{a}{u} \cdot A \cdot J_1\left(\dfrac{u}{a} r\right) \\ H_\theta = 0 \\ H_z = A \cdot J_0\left(\dfrac{u}{a} r\right) \end{cases} \quad \text{for } 0 \le r \le a \tag{2.68a}$$

$$\begin{cases} E_r = 0 \\ E_\theta = i \cdot \omega \cdot \dfrac{a}{w} \cdot \dfrac{J_0(u)}{K_0(w)} \cdot A \cdot K_1\left(\dfrac{w}{a} r\right) \\ E_z = 0 \\ H_r = -i \cdot \beta \cdot \dfrac{a}{w} \cdot \dfrac{J_0(u)}{K_0(w)} \cdot A \cdot K_1\left(\dfrac{w}{a} r\right) \\ H_\theta = 0 \\ H_z = \dfrac{J_0(u)}{K_0(w)} \cdot A \cdot K_0\left(\dfrac{w}{a} r\right) \end{cases} \quad \text{for } a < r. \tag{2.68b}$$

where $u = a \cdot \kappa = a\sqrt{k^2 n_1^2 - \beta^2}$, $w = a \cdot \sigma = a \cdot \sqrt{\beta^2 - k^2 n_0^2}$. Expressions (2.67) and (2.68) give the complete solution for the propagation constants and fields of TE modes in the step-index fiber case.

2.7.2 TM Modes

The dispersion relation for TM modes has the form [5]

$$\frac{J_1(u)}{u \cdot J_0(u)} = -\left(\frac{n_0}{n_1}\right)^2 \frac{K_1(w)}{w \cdot K_0(w)} \tag{2.69}$$

while the electromagnetic fields for the TM mode are

$$\begin{cases} E_r = i \cdot \beta \cdot \frac{a}{u} \cdot A \cdot J_1\left(\frac{u}{a} r\right) \\ E_\theta = 0 \\ E_z = A \cdot J_0\left(\frac{u}{a} r\right) \\ H_r = 0 \\ H_\theta = i \cdot \omega \cdot n_1^2 \cdot A \cdot J_1\left(\frac{u}{a} r\right) \\ H_z = 0 \end{cases} \quad \text{for } 0 \le r \le a \tag{2.70a}$$

$$\begin{cases} E_r = -i \cdot \beta \cdot \frac{a}{w} \cdot \frac{J_0(u)}{K_0(w)} \cdot A \cdot K_1\left(\frac{w}{a} r\right) \\ E_\theta = 0 \\ E_z = \frac{J_0(u)}{K_0(w)} \cdot A \cdot K_0\left(\frac{w}{a} r\right) \\ H_r = 0 \\ H_\theta = -i \cdot \omega \cdot n_0^2 \cdot \frac{a}{w} \cdot \frac{J_0(u)}{K_0(w)} \cdot A \cdot K_1\left(\frac{w}{a} r\right) \\ H_z = 0 \end{cases} \quad \text{for } a < r. \tag{2.70b}$$

The constant A can be determined from the total power carried in optical fiber using the following equation:

$$P = \frac{\pi}{2} \cdot \omega \cdot \beta \cdot |A|^2 \cdot \frac{a^4 v^2}{u^4} \cdot J_1^2(u) \cdot \frac{K_0(w) \cdot K_2(w)}{K_1^2(w)} \tag{2.71}$$

where v is a normalized frequency, defined as

$$u^2 + w^2 = k^2 (n_1^2 - n_0^2)\, a^2 = v^2$$

As an illustration let us consider the axially symmetrical step-index optical fiber at 1550 nm wavelength having $n_0 = 1.44$ and $n_1 = 1.446$ (refractive index difference of 0.4%, which is typical for single-mode optical communications fibers). The dispersion curves for such fibers calculated according to (2.67) and (2.69), are given in figure 2.15.

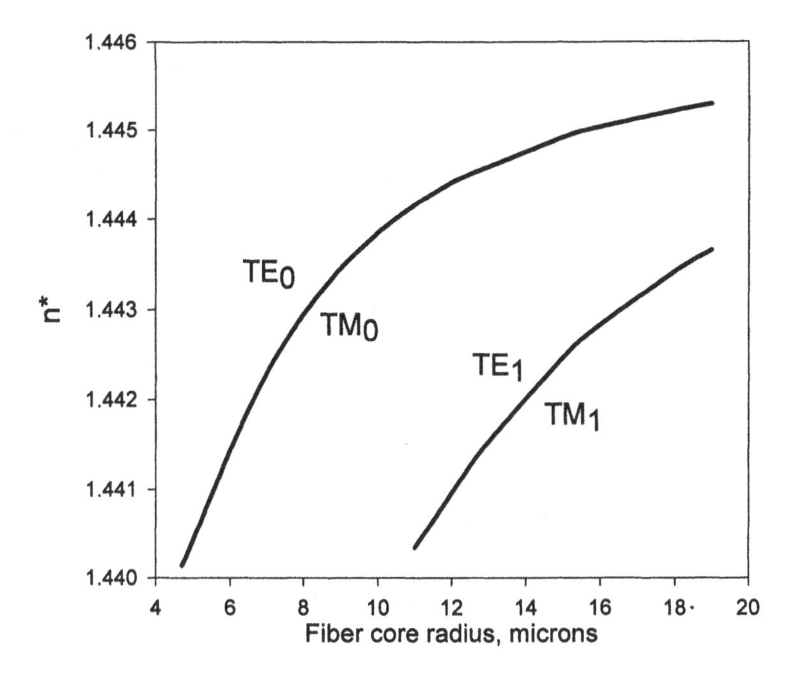

Figure 2.15. Dispersion curves for axially symmetrical step-index optical fiber.

One can see that, in order to have such a fiber single-mode at 1550 nm wavelength, the core diameter has to be between ~ 8 and 20 microns. In figure 2.16 the electric field distribution across the fiber cross-section is given for the first two TE waveguide modes.

2.8 Leaky Dielectric Waveguides

Section 2.6 discussed the properties of TE and TM modes propagating in thin film dielectric waveguides. It was shown that in order to obtain lossless propagation through the dielectric slab, the index of refraction of the waveguiding layer (core) should exceed the refractive indices of the media on both sides of said waveguiding layer. This is a necessary condition for optical waveguiding, although it is not sufficient; the thickness of the waveguiding layer should exceed the cut-off thickness. However, guiding light with very small losses is possible if the core refractive index is lower than that of at least one medium surrounding the waveguiding layer. In this case, total internal reflection, required in order to suppress the reflection losses completely, cannot be achieved at the interfaces. The light wave coupled into such a structure will thus lose power by "leaking" to the bounding medium with a refractive index exceeding that of the waveguiding layer (or into both media, if their refractive indices exceed those of the waveguiding layer). That is why such a structure is called a *leaky waveguide*.

TE0

TE1

Figure 2.16. TE waveguide modes electric field distribution across the fiber cross-section.

Low losses in a leaky waveguide are achieved due to the fact that the Fresnel reflectivity at the interfaces approaches unity when the angle of incidence reaches 90 deg. (see figure 2.2 from section 2.3). Leaky waveguide mode characteristics, such as propagation constants (or effective refractive indices) and loss coefficients, can be derived using the approach taken in section 2.6 for waveguide analyses. The derivation will be omitted here due to its similarity in the two analyses. The TE mode propagation constant, n_{TE}*, for a single-layer leaky waveguide structure with waveguiding layer n_2 having thickness d surrounded by media with refractive indices n_1 and n_3 will be the solution of the analytical equation:

$$\tan(k_{2z}d) = i\frac{k_{2z}(k_{1z} + k_{3z})}{k_{2z}^2 + k_{1z}k_{3z}} \tag{2.72}$$

where $k_{iz} = \frac{\omega}{c}\sqrt{n_i^2 - n_{TE}{}^{*2}}$, $i = 1,2,3$.

Equation (2.72) is very similar to equation (2.57) for waveguide modes. The difference is due to the complex nature of the leaky mode propagation constant: $\beta = (\omega/c)\mathrm{Re}(n_{TE}{}^*) - i\alpha/2$, where $\alpha > 0$ is the *mode power attenuation coefficient*, also known as the *mode loss coefficient*. The complex propagation constant β corresponds to an exponential decay of the leaky waveguide mode power along the mode propagation direction.

As in the waveguide case, there is no analytical solution of (2.72). It can be solved approximately (see [19]) or numerically. To get a feeling for the behavior of the leaky waveguide modes, let us analyze the case of the leaky waveguiding silicon dioxide layer on top of a silicon substrate surrounded by air at 1550 nm wavelength. The numerically calculated dependences of leaky waveguide mode effective refractive index on the silicon dioxide layer thickness are given in figure 2.17a, while the loss coefficients for the same modes are given in figure 2.17b. Only the three first modes of both polarizations are shown. The general behavior of n* for leaky waveguide is similar to the behavior of waveguide n* (see figure 2.9 from section 2.6). The loss coefficients of each mode are decreased with the increase of the silicon dioxide layer thickness. In general, for the analyzed structure, the losses near the leaky waveguide mode cut-off conditions are extremely high: $\alpha = 10000$ cm^{-1} means that the light coupled into such a leaky waveguide would lose 99% of its power on just 10 microns of the leaky waveguide length! However, due to the fast reduction of losses with the silicon dioxide layer thickness, such a structure becomes practical when the silicon dioxide layer thickness exceeds roughly 2.5 microns.

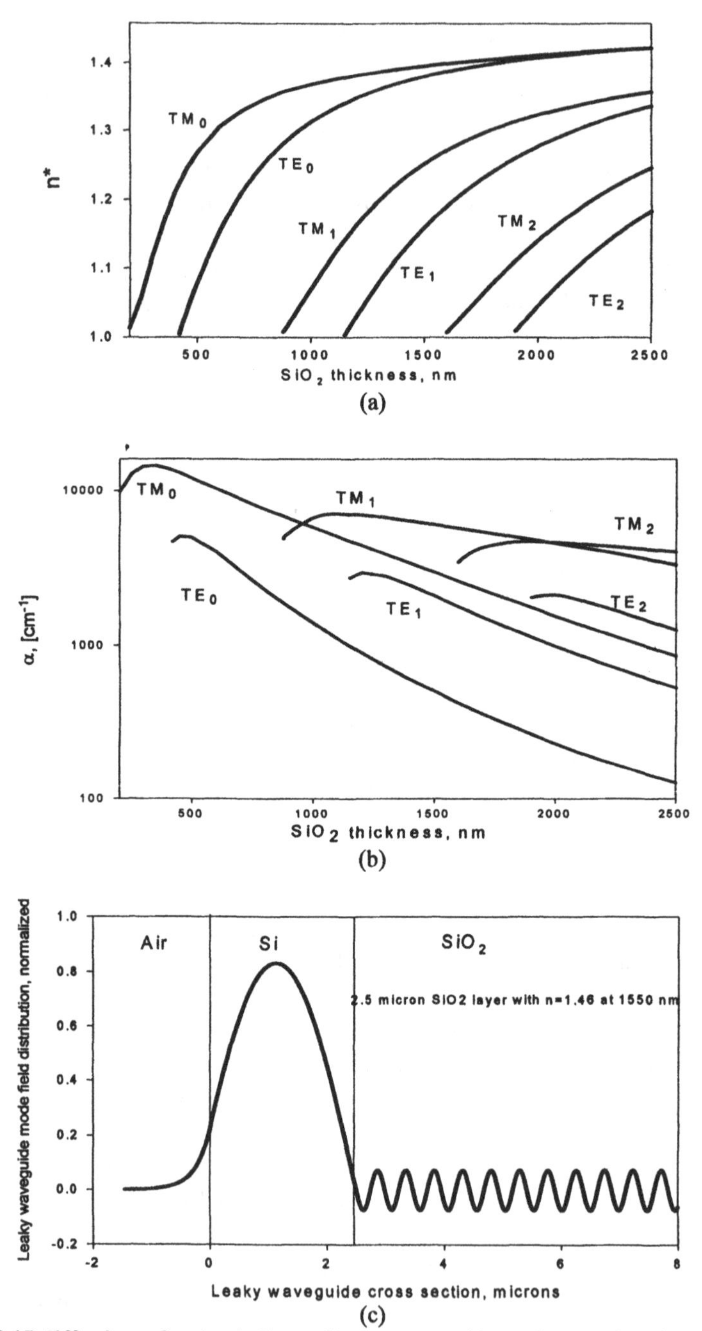

Figure 2.17. Effective refractive indices of leaky waveguide modes as a function of SiO_2 layer thickness for a silica-on-silicon leaky waveguide at 1550 nm wavelength (a); mode loss coefficients as a function of SiO_2 layer thickness (b); and the leaky waveguide TE_0 mode field distribution across the leaky waveguide cross-section for a silica-on-silicon leaky waveguide structure for the thickness of SiO_2 layer of 2.5 microns and the wavelength of 1550 nm.

Figure 2.17c gives the distribution of the TE_0 leaky waveguide mode for 2.5 micron silicon dioxide layer. It is clearly visible that the mode is not confined completely but is leaking to the silicon substrate.

Despite lack of confinement, leaky waveguides have applications in optical cross-connectors, lasers, etc. The reason for this is that, unlike the waveguide case, the light-guiding layer can have a refractive index that is less than that of the substrate, which is important when dealing with high refractive index semiconductor substrates.

2.9 The Bragg Grating

A *fiber Bragg grating* (*FBG*) is a series of fiber areas that create a periodic pattern of differing indices of refraction in the direction of propagation of a light beam (Figure 2.18). A Bragg grating is formed in an optical fiber by exposing a photosensitive fiber core (usually germanium doped fiber) with an ultraviolet (UV) beam that spatially varies periodically in intensity, usually accomplished by means of an interference pattern created by a phase mask or split beam, such as with a Lloyd's mirror apparatus. Light reflections that are caused by the periodic index of refraction pattern in the resulting FBG interfere constructively and destructively similarly to a multilayer interference filter. The refractive index contrast between UV-exposed and unexposed sections of fibers is small (Δn in FBG is usually within the range of 10^{-5} to $2 \cdot 10^{-3}$ [4]). The number of exposed and unexposed sections is very large, so the reflected beam narrows its spectrum to a very sharp peak, as narrow as a fraction of a nanometer in spectral width. FBGs are now being used in a variety of applications such as optical filters for wavelength division multiplexing (WDM) [25], optical sensors [26], dispersion compensation systems [27], and all-fiber lasers.

Modeling of FBG reflection and transmission spectra is possible using the multilayer stack approach described in section 2.5. Two such methods – *effective index method* (*EIM*) [28] and *transfer matrix method* (*TMM*) [29] have been developed. In EIM, the grating is divided into sections, with the length of each one being at least one order of magnitude smaller than the smallest value of the corrugation period. The electromagnetic wave components are calculated in each section under the assumption that the refractive index remains constant in each section. In general this technique is good for complex FBG structures (i.e., when the grating has more than one period, phase shift, etc.); although for the analyses of simple single-period FBGs, it require excessive computation time. For example, for a 2 cm FBG with the period of 538 nm (designed to have peak reflectivity at 1550 nm), EIM method will require computation of reflection and transmission coefficients from at least 370,000 layers! In TMM the FBG is divided into sections, with the length of each section being much bigger than the biggest period of the corrugation. The index variation in each section is considered to be in the form of uniform grating. Each of these sections is calculated as a multilayer stack, and overall transmission and reflection are obtained as a product of individual matrices. In general TMM is a considerably faster approach than EIM; although it is not as straightforward for complex FBG profiles.

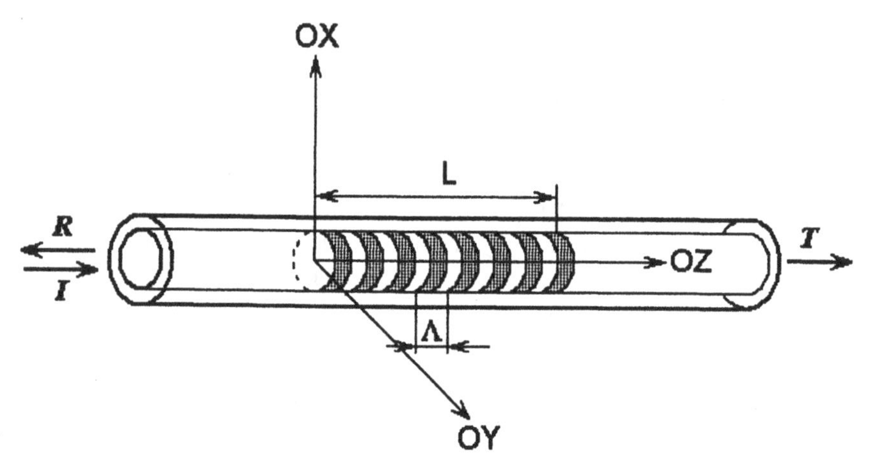

Figure 2.18. A schematic drawing of an intracore FBG. The coordinates system, used in the text, is shown along with the regions of changed refractive index.

The fastest method for analyzing FBGs is the *coupled-mode method* (*CMM*). It combines simplicity, accuracy, and fast speed in modeling the optical properties of most FBGs. This approach is especially useful for describing simple, one-period uniform FBGs. The derivation of CMM can be found in, for example, [30] or [31].

The derivation of CMM presented here closely follows the work of [31]. The important assumption here is that the perturbation of the fiber modes by the inscribed FBG is so small that it can be neglected. Under this assumption, the transverse component of the electric field in the fiber within the FBG can be presented as a sum of the transverse components of each mode, supported by the fiber:

$$E_T(r,z,t) = \sum_i \left\{ A_i(z) e^{i \cdot \beta_i \cdot z} + B_i(z) e^{-i \cdot \beta_i \cdot z} \right\} \mathrm{E}_i(r) e^{-i \cdot \omega \cdot t}$$

where i is the mode label, A_i(z) is forward-propagating (+z direction in the figure 2.18) i^{th} mode amplitude; B_i(z) is backward propagating (-z direction in the figure 2.18) i^{th} mode amplitude; E_i(r) is the transverse field of ith mode (cylindrical symmetry of the fiber is assumed); ω is the frequency of light,

$\beta_i = (2\pi/\lambda)n_i^*$ is the propagation constant of ith mode. Under the above assumptions, the modes are orthogonal, i.e., there is no mechanism is available for energy exchange (coupling) between them other than through inscribed grating. For the simplification of the formulas, the most important case of the coupling of counter-propagating modes in single-mode fiber will be described here. In this case the amplitudes of counter-propagating modes are connected by the following equation [31]:

$$\left.\begin{aligned}\frac{dA(z)}{dz} &= i\cdot k\cdot B(z)\cdot e^{-i\cdot 2\Delta\beta\cdot z}\\ \frac{dB(z)}{dz} &= -i\cdot k^*\cdot A(z)\cdot e^{i\cdot 2\Delta\beta\cdot z}\end{aligned}\right\} \tag{2.73}$$

where k is the mode coupling coefficient. For the single-harmonic refractive index modulation in FBG, the resultant perturbation of index of refraction n^* is given by the following expression:

$$\delta n^*(r,z) = \overline{\delta n^*}\left(1 + s\cos\left(\frac{2\pi}{\Lambda}z\right)\right)$$

where s is the fringe visibility associated with the index change and $\overline{\delta n(r,z)}$ is the "dc" index change spatially averaged over the grating period. For most types of photosensitive fiber currently in use the UV-induced refractive index changes $\overline{\delta n(r,z)}$ are approximately uniform across the fiber core and negligible outside the core. In this case the mode-coupling coefficient can be determined as

$$k = k^* = \frac{\omega\cdot n_{core}\cdot\overline{\delta n}_{core}\cdot s}{4}\int_{core}\mathrm{E}(r)\cdot\mathrm{E}(r)^*\,dr = \frac{\pi\cdot\overline{\delta n}^*\cdot s}{\lambda}$$

The boundary conditions should be set as $A(0) = \sqrt{I_0}$ (where I_0 is the intensity of the wave incident of FBG); $B(L) = 0$ where L is FBG length (i.e., no reflected wave exists at the output of an FBG section [31]). The solution of (2.73) in this case for a forward wave E_i and backward wave E_r will be

$$E_i(z) = \sqrt{I_0}\,\frac{\{\gamma\cdot\cosh[\gamma\cdot(z-L)] - i\cdot\Delta\beta\cdot\sinh[\gamma\cdot(z-L)]\}\cdot e^{i\beta_0\cdot z}}{i\cdot\Delta\beta\cdot\sinh(\gamma\cdot L) + \gamma\cosh(\gamma\cdot L)} \tag{2.74a}$$

$$E_r(z) = \sqrt{I_0}\,\frac{i\cdot k\cdot\sinh[\gamma\cdot(z-L)]\cdot e^{-i\beta_0\cdot z}}{i\cdot\Delta\beta\cdot\sinh(\gamma\cdot L) + \gamma\cosh(\gamma\cdot L)} \tag{2.74b}$$

where $\gamma^2 = |k|^2 - (\Delta\beta)^2$; $\Delta\beta = \beta - \beta_0$, where $\beta_0 = \pi l/\Lambda$, Λ is a period of Bragg grating, and l is the integer number corresponded to the diffraction order of FBG, $\beta = \beta(\lambda)$ is a propagation constant of the unperturbed fiber. The complex transmission coefficient t through whole FBG would be in this case $t = E_t(L)/\sqrt{I_0}$ and the complex reflection coefficient r will be $r = E_r(0)/\sqrt{I_0}$:

$$t = \frac{\gamma \cdot e^{i\beta_0 \cdot L}}{i \cdot \Delta\beta \cdot \sinh(\gamma \cdot L) + \gamma \cosh(\gamma \cdot L)} \qquad r = \frac{-k \cdot \sinh(\gamma \cdot L)}{i \cdot \Delta\beta \sinh(\gamma \cdot L) + \gamma \cosh(\gamma \cdot L)}$$

From here the reflectance $R = r \cdot r^*$ and the transmittance $T = 1 - R$ (if optical losses in the fiber are neglected) can be found.

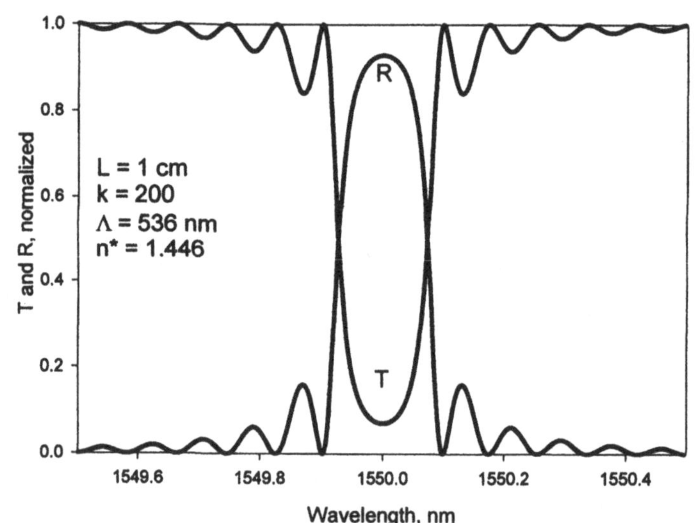

Figure 2.19. Calculated transmission and reflection characteristics of FBG.

Figure 2.19 shows the numerically calculated (using CMM) transmittance and reflection through the FBG with the length of 2 cm, mode coupling coefficient k of 200 cm^{-1}, period Λ = 536 nm written into a typical optical fiber with n^* = 1.446 at 1550 nm wavelength. The FBG acts as a band-blocking filter that reflects the light near the Bragg wavelength $\lambda_B = 2n^*\Lambda$, and transmits light outside the Bragg wavelength. The angular frequency width of the FBG reflection band is proportional to the mode-coupling coefficient in FBG k: $\Delta\omega_B = \dfrac{2k \cdot c}{n^*}$, which in turn is proportional to the value of the refractive index modulation in FBG. In other words, the stronger the coupling between the modes in FBG, the wider the FBG reflection wavelength range. The maximum reflection efficiency of FBG is achieved at the Bragg wavelength λ_B and is equal to $R = \tanh^2(k \cdot L)$. The experimentally measured transmission spectrum of the typical FBG is given in figure 2.20.

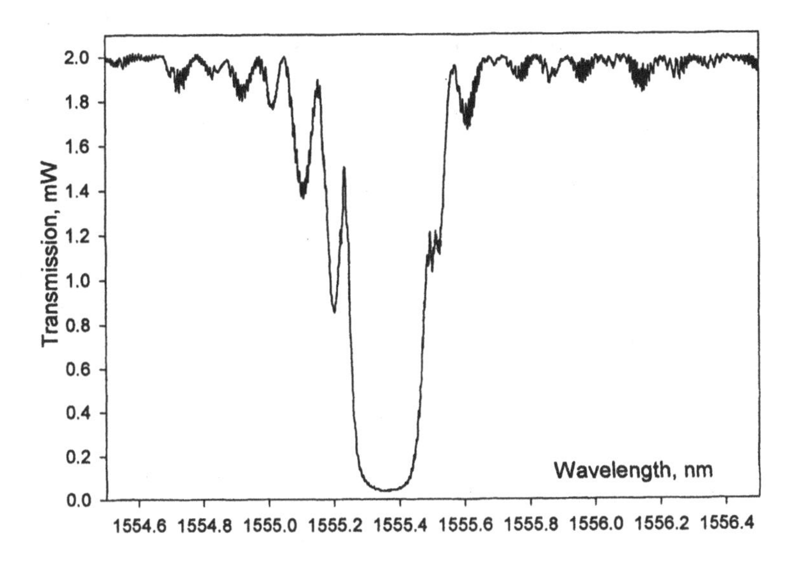

Figure 2.20. Experimental transmission spectrum of commercially available FBG.

The experimental spectra of FBGs are usually more complex structurally and are frequently asymmetrical. It happens due to nonideal manufacturing conditions, such as nonuniform exposure over the length of the FBG.

The uniform single-period FBG is not optimized for optical filter applications due to presence of strong side-lobes and not very sharp blocking edges (see figures 2.19 and 2.20). Because of that, FBGs used for WDM filters usually have a more complex structure than just simple one-period uniform FBGs. Several ways exist to overcome these difficulties. The most frequently used methods are apodization of FBGs [32] and use of multiple phase shifts in FBG similar to the usual multilayer filters (which are discussed in the next chapter). The discussion of the many different types and structures of FBGs; however is beyond the scope of this book.

2.10 Codirectional coupling

Previous sections discussed the transmission and reflection characteristics of independent optical waveguides. To understand the material presented in the following chapters of this book, it is important to consider the interaction between the two copropagating light beams in the adjacent waveguides – so-called codirectional coupling. The most frequently used method for such a problem is the *coupled mode method* (*CMM*) (see [5], [19]). CMM is based on perturbation theory and possesses the derivation of elegant analytical expressions for this problem. However, this problem will be analyzed here by interference phenomena between even and odd modes in two parallel waveguides separated by a finite distance [20]. The author believes that this computational approach is preferable from the viewpoint of simplicity and the potential of use in commercially available software.

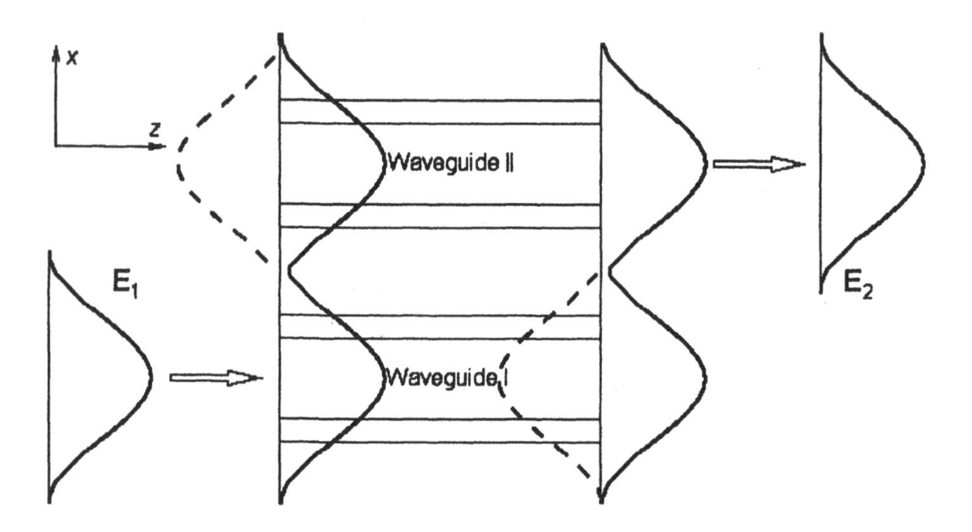

Figure 2.21. Even (solid line) and odd (dashed line) modes in two coupled multilayer waveguides.

Only a one-dimensional case will be considered here. The two-dimensional cases are similar to one-dimensional case, although they require more sophisticated computation. The geometry of the problem under consideration is shown in figure 2.21: two multilayer planar waveguides, denoted as "Waveguide I" and "Waveguide II" are spaced apart by some distance. For the sake of simplicity, both waveguides are assumed to be single-mode. $E_I(x)$ denotes the mode field distribution of waveguide I while $E_2(x)$ denotes the same for waveguide II. In this case the electric field in such a system of two adjacent waveguides can be approximated by summation of the even mode (first-order mode in the multilayer waveguide, obtained by placing two initial waveguides close together) and the odd mode (second-order mode):

$$E(x,z) = E_e(x)\exp(-i2\pi n_e^{*}\cdot z/\lambda) + E_o(x)\exp(-i2\pi n_o^{*}\cdot z/\lambda) \quad (2.75)$$

Where $E_e(x)$ and n_e^* denote the electric field and effective refractive index of the even mode and $E_o(x)$ and n_o^* denote those of the odd mode, respectively.

The incident electric field E_1 is assumed to be coupled at $z = 0$ to waveguide I only, i.e.,

$$|E(x,0)| = |E_e(x) + E_o(x)| = E_1(x)$$

This expression provides good accuracy when the distance between waveguides I and II is sufficient to neglect the individual modes perturbation in the waveguides. In this case, expression (2.75) can be rewritten as

$$|E(x,z)| = |E_e(x) + E_o(x)\exp(i2\pi[n_e^* - n_o^*]\cdot z/\lambda)|$$

Electric field distribution at $z = \lambda/2(n_e^* - n_o^*)$ then will be

$$|E(x, \lambda/2[n_e^* - n_o^*])| = |E_e(x) - E_o(x)| = E_2(x)$$

This means that the incident field coupled to waveguide I shifted to waveguide II at the distance

$$L_c = \lambda/2(n_e^* - n_o^*) \qquad (2.76)$$

L_c is a coupling length. Due to the time invariance of the Maxwell equations, result will occure the same if the incident field is coupled at $z = 0$ to waveguide II: in this case at $z = L_c$ the electric field will shift to the waveguide I.

To illustrate this, let us consider the following example: two waveguide cores with refractive indices 1.442 and thickness 8.5 microns are embedded in the media with refractive index 1.44. The task is to find the dependence of coupling length L_c on the waveguide core separation at the wavelength of 1550 nm. In figure 2.22, the electric field distributions along the x-axis (see figure 2.21 above) are presented for the even and odd modes of such a system for 5 microns separation between cores together with mode fields of the same cores if the second core is missed. For this particular separation of the cores, the perturbation of the mode field in each waveguide caused by the coupling to another waveguide is quite strong and the accuracy of this approach is rather poor.

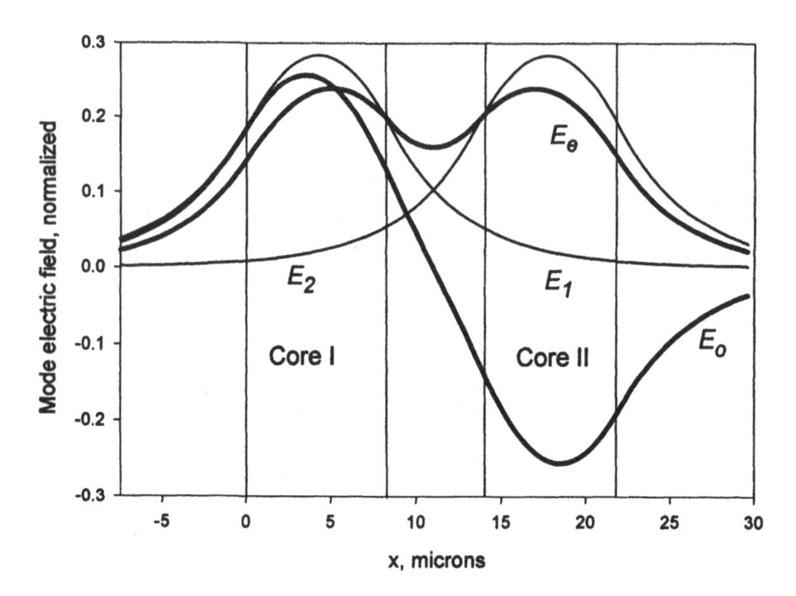

Figure 2.22 Electric field distributions along the x-axis for the even and odd modes for 5 microns separation.

Figure 2.23 gives the numerically calculated effective refractive indices of even and odd modes as functions of waveguide core separation (a) and the coupling length L_c recalculated according to (2.76) is given as functions of waveguide core separation (b). As expected, the coupling length increased with the waveguide core separation exponentially (note the logarithmic scale of L_c in figure 2.23b).

This directional coupler found wide use in optical communications due to the relative simplicity of manufacturing.

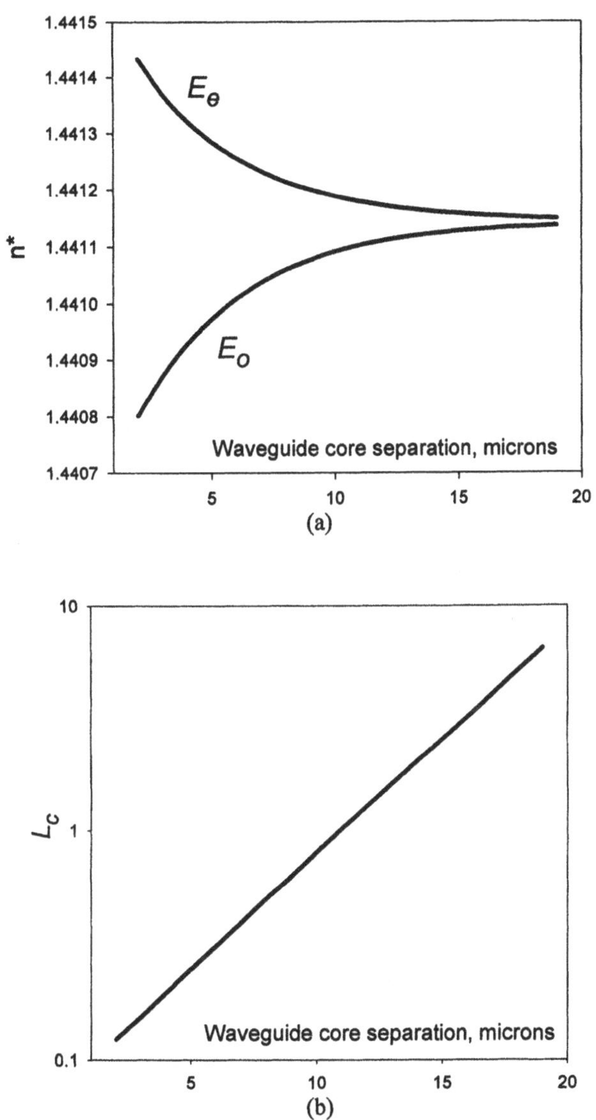

Figure 2.23. Dependences of the effective refractive indices of even and odd modes on a waveguide core separation (a) and the dependence of the coupling length L_c on a waveguide core separation (b) for a waveguide structure described in the text.

Chapter 3

CLASSICAL OPTICAL FILTERS

3.1 Antireflection Coatings

As was demonstrated in section 2.3 of this book, the reflectance from an air/glass (or a glass/air) boundary is about 4% through the visible wavelength range at normal incidence. Hence, each glass optical element (lenses, waveplates, etc.) working at transmission mode is responsible for at least 8% of the losses in transmitted power. The situation is even worse for optical elements made of higher refractive index materials. Therefore, the total transmittance of even relatively simple optical systems can be quite low. In addition, light reflected from the various surfaces in the optical system can (and usually does) reach the focal plane, which leads to the degradation of optical system performance. To solve these problems, *antireflection coatings* are used. Antireflection coating is usually understood to be the coating of the optical surface minimizes the reflection from the surface at some angles of incidence over a specified wavelength range.

The simplest type of antireflection coating is *single-layer antireflection coating* or shorter *antireflection layer* (*ARL*).

3.1.1 Single-Layer Antireflection Coatings

To analyze single-layer antireflection coating, it is worthwhile recalling the consideration of thin film optics, given in section 2.4 of this book. The three-layer optical system that we will consider was shown in the figure 2.4. Here we will use the abbreviations presented in section 2.4: medium 0 is (as before) the incident medium, medium 2 is the substrate, and layer 1 is the antireflection layer, whose parameters should be chosen to minimize the reflection from such a three-layer system. The general expressions for reflection coefficients for such a system were found to be (2.40), (2.41), and (2.46) for TM modes and (2.44) to (2.46) for TE modes.

Let us define optical thickness of the antireflection layer as

$$D = n_1 d \tag{3.1}$$

where θ_1 is the angle between the direction of propagation of electromagnetic wave inside the antireflection layer and the normal direction to the surfaces between media. In this case the condition for layer 1 to be the antireflection layer for such an optical system would be

$$\begin{cases} \dfrac{dR_{02}}{dH} = 0 \\ \dfrac{d^2 R_{02}}{dH^2} > 0 \end{cases} \tag{3.2}$$

From (2.40) and (2.41) and (2.44) to (2.46), by using elementary math, we can find

$$R_{02} = |r_{02}|^2 = \left| \frac{r_{01} + r_{12} e^{2i \cdot k_{1z} d}}{1 + r_{02} r_{02} e^{2i \cdot k_{1z} d}} \right|^2 = \\ = \frac{r_{01}^2 + r_{12}^2 + 2 r_{01} r_{12} \cos 2k_{1z} d}{1 + r_{01}^2 r_{12}^2 + 2 r_{01} r_{12} \cos 2k_{1z} d} \tag{3.3}$$

By recalling the definition of k_{1z} from section 2.3.1, we can obtain

$$k_{1z} = \sqrt{\varepsilon_1 \left(\frac{\omega}{c}\right)^2 - k_x^2} = \sqrt{n_1^2 \left(\frac{\omega}{c}\right)^2 - n_0^2 \left(\frac{\omega}{c}\right)^2 \sin^2 \theta_0} = \frac{\omega}{c} n_1 \cos \theta_1$$

and $(\omega/c) = (2\pi/\lambda)$, where λ is the wavelength of monochromatic electromagnetic wave. In this case (3.3) can be rewritten as

$$R_{02} = \frac{r_{01}^2 + r_{12}^2 + 2 r_{01} r_{12} \cos\left(2 \dfrac{2\pi}{\lambda} D \cos \theta_1\right)}{1 + r_{01}^2 r_{12}^2 + 2 r_{01} r_{12} \cos\left(2 \dfrac{2\pi}{\lambda} D \cos \theta_1\right)} \tag{3.4}$$

By using elementary math, from (3.2) and (3.4) we can find that the reflectance R_{02} reaches minimum value at the following conditions:

$$D = n_1 d = \frac{m\lambda}{4 \cos \theta_1}, \; m = 1,3,5,\ldots \tag{3.5}$$

At the conditions of (3.5) the reflectivity (3.4) becomes

$$R_{02} = \left(\frac{r_{01} - r_{12}}{1 - r_{01} r_{12}} \right)^2 \tag{3.6}$$

For the case of normal incidence, according to (2.41) or (2.45), expression (3.6) takes the form

$$R_{02} = \left(\frac{n_0 n_2 - n_1^2}{n_0 n_2 + n_1^2} \right)^2 \qquad (3.7)$$

Expressions (3.5) and (3.7) define both the optimal refractive index of the antireflective layer and its optimal thickness, which for the normal incidence takes the following values:

$$\begin{cases} n_{ARL} = \sqrt{n_0 n_2} \\ d_{ARL} = \dfrac{m\lambda}{4n_{ARL}}, m = 1,3,5,... \end{cases} \qquad (3.8)$$

Under the conditions (3.8), the reflectance from the substrate becomes zero at normal incidence.

The most frequently used case is when the incident medium is air with the index of refraction being very close to unity. In this case n_{ARL} should be equal to $(n_s)^{1/2}$ (where s denotes "substrate") to get the optimal ARL performance. Unfortunately, we do not have an unlimited choice of available refractive indices for the ARL. So in practice, one usually chooses the material for the ARL with the value closest to $(n_s)^{1/2}$. Another important requirement for the ARL is that optical absorption of the ARL should be zero or very small to minimize losses in the ARL.

Such an approach works best with high-refractive index substrates, for example, semiconductors. As an example, let us consider an ARL for a silicon substrate at a wavelength of 1550 nm that is important for optical communications. At this wavelength the refractive index of silicon is around 3.5. Hence, the ARL refractive index should be around $(3.5)^{1/2} = 1.871$. In the classical paper [33], silicon monoxide was recommended as the ARL material for silicon, since it has a refractive index that is close to the optimum $n_{\mathrm{SiO}} = 1.9$ and exhibits very low losses in the desired wavelength range.

Figure 3.1 shows the angular dependences of the silicon, silicon coated by 204 nm of silicon monoxide (n = 1.9), and silicon coated by 269 nm of silicon dioxide at 1550 nm wavelength. The decrease in reflectivity is significant for the ARL from SiO_2, even though the refractive index of silicon dioxide is 23% smaller than the desired value of 1.871.

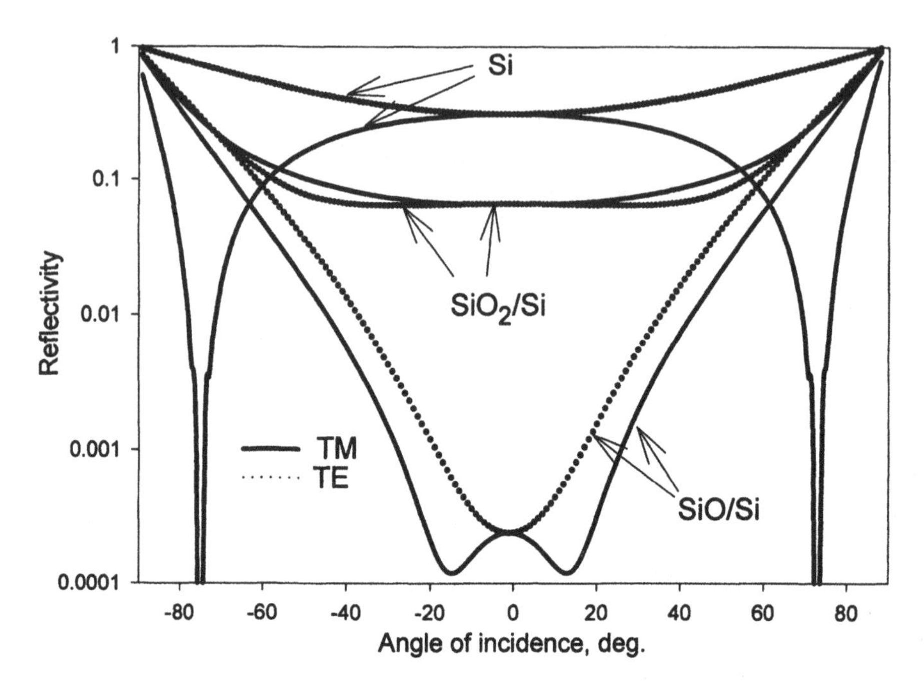

Figure 3.1. Angular reflectivity spectra for silicon, silicon coated by 204 nm of silicon monoxide (n = 1.9), and silicon coated by 269 nm of silicon dioxide at 1550 nm wavelength.

For silicon monoxide the decrease in reflectivity is dramatic– three orders of magnitude. The reflectivity of the silicon monoxide-coated silicon wafer with the right thickness of silicon monoxide can be considered to be zero for most applications. Figure 3.1 also illustrates a typical disadvantage of single-layer antireflection coatings: according to (3.5) it is possible to get optimized performance of the ARL only for some relatively narrow range of angles of incidence. The performance of the silicon monoxide ARL on silicon substrate having thickness optimized for normal incidence degrades by an order of magnitude at angles of incidence around 30 deg. Also, the ARL performance at nonzero angles of incidence exhibits relatively strong polarization dependence. These disadvantages lead to the necessary modifying the ARL parameters depending on the geometry, in which the substrate will be oriented with respect to the incident light.

Another drawback of the single-layer antireflection coatings is that according to (3.5) and (3.8), the ARL can be optimized only for a relatively narrow wavelength range. Figure 3.2 presents the reflectivity spectra of silicon, coated by 204 nm of silicon monoxide (n = 1.9), for different angles of incidence. The performance of such an ARL degrades by more than an order in magnitude at the wavelengths about 10% apart from the designed wavelength. With the increase of the angle of incidence according to (3.5), the reflectivity deep shifts toward lower wavelengths and becomes less pronounced. The position of the deep is the same for both TM and TE polarizations of incident light. However, the minimal achievable reflectivities are quite different for different polarizations.

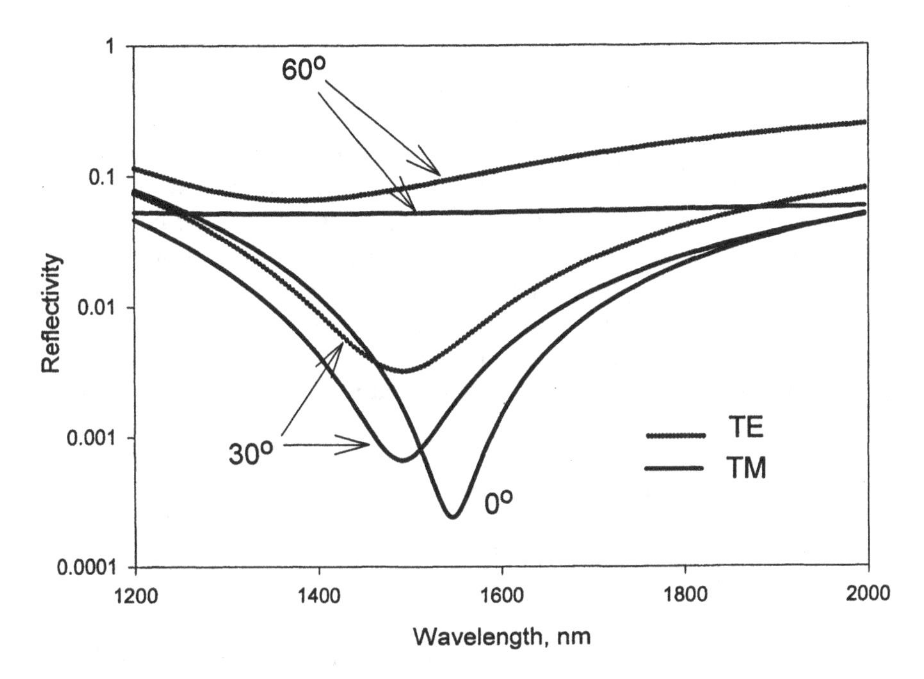

Figure 3.2. Reflectivity spectra of silicon coated by 204 nm of silicon monoxide (n = 1.9) at different angles of incidence.

3.1.2 Multilayer Antireflection Coatings

The above section discussed the single-layer antireflection coating. As mentioned there, such a coating can be sufficient for monochromatic light beams with relatively low divergence (or convergence), but is quite a poor performer for applications, which require polychromatic beams with high divergence (or convergence). The typical example of such applications is the optical microscope. Another very important problem with single-layer antireflection coatings arises when dealing with relatively low refractive index substrates like glass or quartz. The optimum refractive index of ARL layer for an air/glass interface at 1550 nm wavelength will be $(1.44)^{1/2} = 1.2$ (see expression (3.8)). Unfortunately, there are no solid materials with such a refractive index. The lowest useful film index that can be deposited, sputtered or grown is that of magnesium fluoride, about 1.38 at 500 to 600 nm wavelengths. Hence, single-layer antireflection coatings are not suitable for low refractive index substrates. The classic way to overcome such a problem is to use *multilayer antireflection coatings*.

A detailed description of multilayer antireflection coating designs is beyond the scope of this book. Readers can find an excellent description of this topic in the classic book by Macleod [34]. Here I give only a brief explanation of the topic, enough to understand the remaining chapters.

There is a wide variety of multilayer antireflection coating designs [34]. They differ by the number of layers used, by particular requirements for the reflection wavelength, angular spectra, and so on. The number of layers used varies from just two layers (which is enough to get reasonable suppression of the reflection from a low refractive index substrate for narrow wavelength and angles of incidence ranges) to eight or more layers (which are needed to get reflection suppression over quite a wide wavelength and angles of incidence ranges). Also, multilayer antireflection coatings can be designed to suppress the reflection within some wavelength and angular range (like coatings on microscope lenses) or to suppress the reflection at some distinct wavelengths, spaced far from each other (like coatings for multiple frequency laser applications).

There is no analytical design method for all multilayer antireflection coatings. While widely accepted methods exist for two-, three-, and four-layer antireflection coatings, in general, the multilayer antireflection coating design is based mostly on trial and errors.

In general, the thickness of the layers in multilayer antireflection coatings does not have to be quarter wave ($nd = \dfrac{m\lambda}{4\cos\theta}$, $m = 1,3,5,\ldots$). Moreover, the multilayer can consist of a number of different materials. However, the most frequently used multilayer antireflection coatings consist of alternative layers of high refractive index (in the future consideration it will be denoted as H) and low refractive index (in the future consideration it will be denoted as L) materials. The antireflection coating structure can be written in schematic form, for example, {Air | $L\ H\ L\ H$ | Glass} or {Air | $L\ H\ L\ H\ L$ | Glass}.

As an example, let us consider a five-layer antireflection coating on glass that is designed to have a center wavelength around 600 nm. Glass at such a wavelength has the refractive index of 1.52. The low refractive index material in antireflective multilayer will be magnesium fluoride, and the high refractive index material will be titanium dioxide. Figure 3.3 gives the numerically calculated reflectivity spectra for different angles of incidence. The multilayer structure was taken from [34]: $n_0 = 1$; $n_1 = 1.38$ (magnesium fluoride); $d_1 = 126.643$ nm; $n_2 = 2.30$ (titanium dioxide); $d_2 = 41.377$ nm; $n_3 = 1.38$; $d_3 = 47.483$ nm; $n_4 = 2.30$; $d_4 = 27.230$ nm; $n_5 = 1.38$; $d_5 = 246.926$ nm; $n_6 = 1.52$.

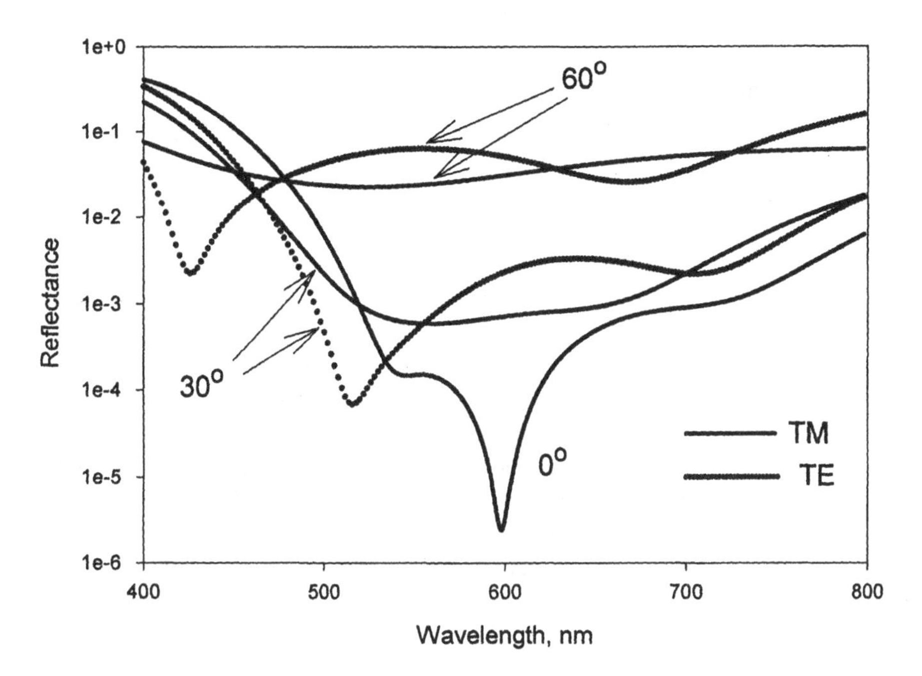

Figure 3.3. Reflectivity spectra of glass coated by five-layer antireflection coating at different angles of incidence.

As follows from figure 3.3, the suppression of the reflectance at the central wavelength of such an antireflection structure is very strong. What is more important is that the reflection at the normal incidence suppressed to below 1% at wide wavelength range – from about 500 nm to more than 800 nm. That is, the width of reflection valley exceeds 50% of the central wavelength (compare to less than 20% for the single-layer ARL of figure 3.2). The angular dependence of such an antireflection coating is also considerably better than that of a single-layer ARL: for 600 nm wavelength, reflection exceeds 1% for angles greater than 40 to 45 degrees depending on polarization of incident light.

Although multilayer antireflection coatings still exhibit angular dependence and are still limited in useful wavelength range, better fit practical needs than single-layer ARL, and their performance is suitable for most applications.

3.2 Neutral Mirrors

Optical mirrors can be considered a type of optical filters that is quite different from antireflection coatings. The main purpose of mirrors is to reflect the incident wave as well as possible. Depending on the application, additional features can be added to high reflection: for high power lasers, for example, the absorption of light should be minimized.

The simplest, oldest and most frequently used type of mirror is the metallic mirror. In this type of mirror a metallic layer provides the reflection. Numerically calculated (according to the formulas (2.32) and (2.34)) the reflectance spectra of freshly deposited films of silver, gold, and copper (refractive indices were taken from [35]) at normal incidence are given in Figure 3.4. These metals are all quite good performers in the infrared and near infrared wavelengths. However, the reflectance of copper drops significantly around 600 nm; the reflectance of gold drops at 550 nm, and the reflectance of silver drops at 360 nm. Aluminum (not shown in the plot) provides reasonably good reflectance down to 200 nm; although at near infrared and red wavelengths, its performance as a mirror layer is worse than that of silver, gold, and copper.

The important advantage of metallic mirrors is the small dependence of the reflectance on the angle of incidence. For example, figure 3.5 shows the angular dependence of the reflectance of silver film at 632.8 nm wavelength. At this wavelength the variations of the reflection for TM polarized waves does not exceed 3%, while for TE polarized waves these variations are even smaller – around 1.5%.

Despite high reflectivity, omnidirectionality, and simplicity of fabrication, metallic mirrors have a significant disadvantage: the absorption. The numerically calculated (according to formulas (2.32), (2.34) and (2.47)) absorptance spectra of freshly deposited films of silver, gold, and copper (refractive indices were taken from [35]) at normal incidence are given in figure 3.6. Even at IR wavelengths absorption is significant for high-power laser applications like Nd:YAG laser (which can provide the optical power on the order of hundreds of Watts). For example, if a silver mirror is illuminated by 100 W Nd:YAG laser at 1064 nm wavelength and 45 degrees incidence, it will absorb 300mW. This is enough to damage the mirror. At visible wavelength range, the situation is even worse (note the logarithmic scale of absorptance in the figure 3.6).

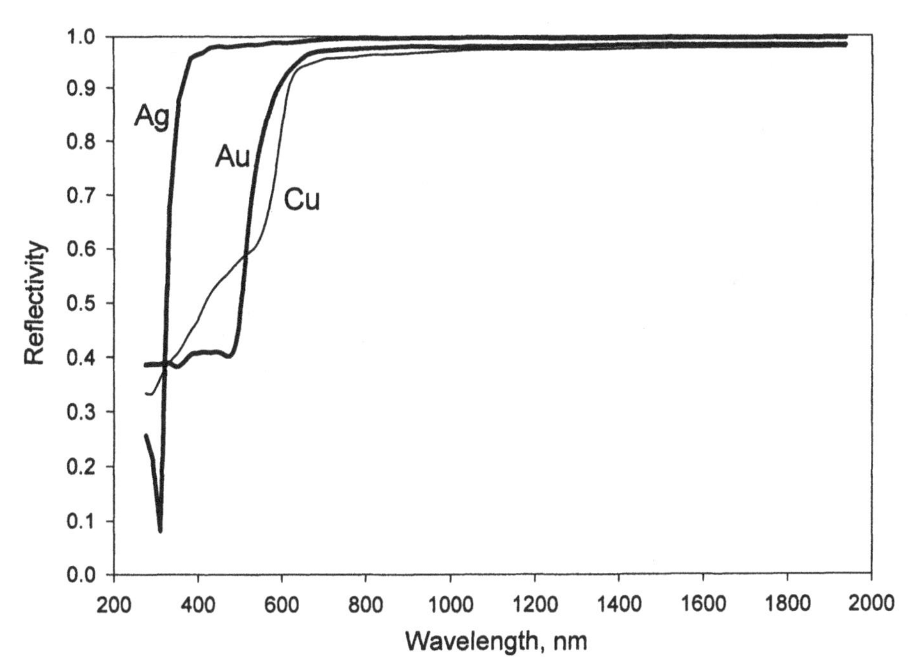

Figure 3.4. Reflectance spectra of freshly deposited films of silver, gold, and copper at normal incidence.

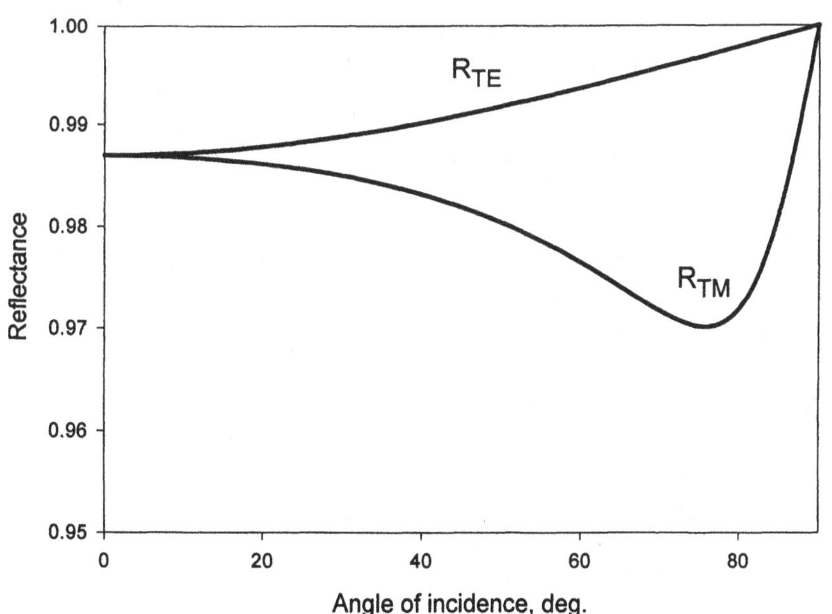

Figure 3.5. Angular dependence of the reflectance of silver film at 632.8 nm wavelength.

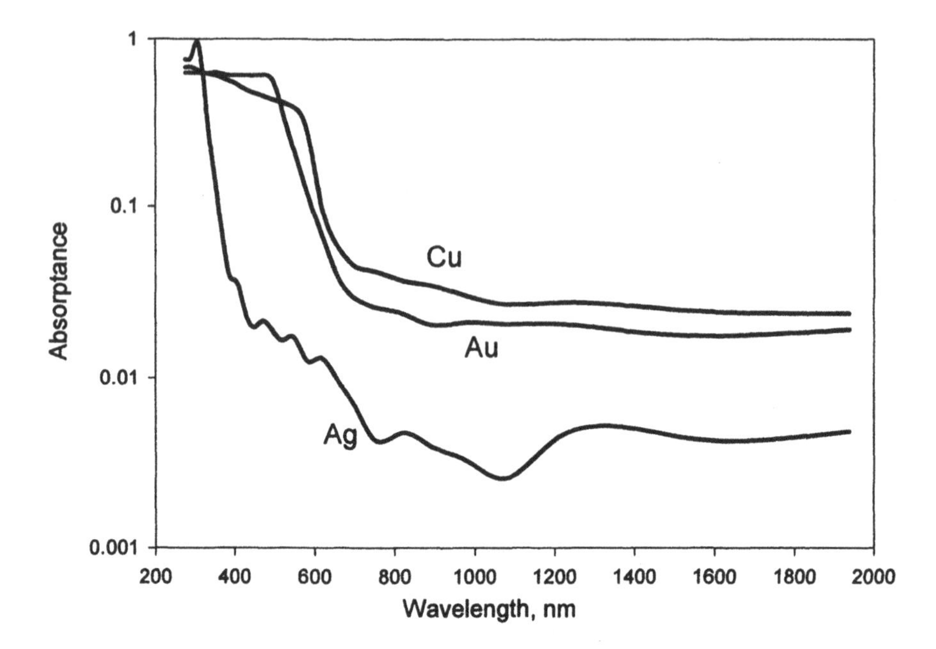

Figure 3.6. Absorptance spectra of freshly deposited films of silver, gold, and copper at normal incidence.

Another problem with metallic mirrors arises at deep UV wavelengths (below ~ 300 nm). As mentioned above, the reflectivity of metals at such wavelengths is poor for gold, silver, and copper and is not very good for aluminum. To solve this problem, an additional dielectric layer with a suitable refractive index is usually deposited on the top of the aluminum layer (for more details, see the classic textbook on optical filter by Macleod [34]).

To solve the problem of absorption and poor performance at deep UV a different types of optical filters are usually used – a filters based on multilayer dielectric coatings.

3.3 Multilayer High-Reflectance Coatings

Multilayer high-reflectance dielectric coatings are based on multiple-beam interference in a multilayer stack. To understand this phenomenon better, let us consider the simplest and most successful realization of multiple beam interferometers – the Fabry-Perot (FP) interferometer.

3.3.1 The Fabry-Perot Interferometer

A Fabry-Perot interferometer can be imagined as a thin film of thickness d having two flat surfaces that are parallel to each other and coated with relatively high-reflectance coatings (see figure 3.7). In practice, a different realization of Fabry-Perot interferometer is common – the so-called Fabry-Perot etalon, which consists of two flat plates separated by a distance *d* and aligned parallel to each other with a high degree of accuracy. The separation is usually maintained by a spacer ring made of quartz or Invar, and the inner surfaces of the two plates are usually coated to enhance their reflection.

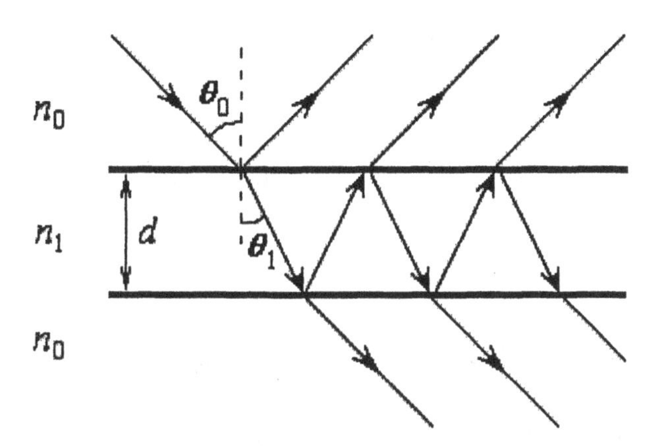

Figure 3.7. A Fabry-Perot etalon.

The analyses of FP interferometer can be done using methodology developed in sections 2.4-2.5 of this book. However, we apply a different approach here (following [18)]), which is somewhat more illustrative and can be used without modifications for both realizations of the Fabry-Perot interferometer mentioned above.

Let us forget for a moment about the detailed structure of reflective coatings of thin-film surfaces (or inner surfaces of plates). Let us assume that the intensity of the wave incident on a Fabry-Perot interferometer is equal to unity. Let us further assume that the complex reflection coefficient for the wave that is reflected at the first surface of thin film is r, while for the wave reflected from the second surface of thin film the coefficient is r', and the same for complex transmission coefficients t and t'. In this case the complex amplitudes of the waves escaping the thin film into the upper medium (medium of incidence) will be (see figure 3.7)

$$r,\ tt'r'\exp\{i\delta\},\ tt'(r')^3\exp\{2i\delta\},\ldots,$$
$$\ldots,\ tt'(r')^{2m-3}\exp\{i(m-1)\delta\},\ldots \qquad (3.9)$$

where $\delta = (4\pi/\lambda)\ n_1 d\ \cos\theta_1$ is the phase shift occurring through double crossing of the thin film (see (3.4)). Analogically, the complex amplitudes of the waves escaping thin film into the bottom medium will be (see figure 3.7)

$$tt',\ tt'(r')^2 \exp\{i\delta\},\ tt'(r')^4 \exp\{2i\delta\},\ \ldots,$$
$$\ldots,\ tt'(r')^{2(m-1)} \exp\{i(m-1)\delta\},\ldots \qquad (3.10)$$

It is follows from (2.40) and (2.41) and from (2.44) to (2.45) (it can be proven that the same is valid for any number of layers) that

$$r = -r';\quad |r|^2 = |r'|^2 = R;\quad |t|^2 = |t'|^2 = T \qquad (3.11)$$

Also, if neither thin-film nor reflective coatings are absorptive, the following is true (see (2.47), which can be proven to be valid for any number of layers):

$$T + R = 1 \qquad (3.12)$$

The amplitude of wave $A^{(r)}$ reflected from Fabry-Perot interferometer will be the sum of all waves escaped into the incident medium. According to (3.9) and using a well-known formula of the sum of geometrical progression one can find

$$A^{(r)} = r + \lim_{m\to\infty} \sum_{k=2}^{m} tt'(r')^{(2k-3)} \exp(i(k-1)\delta) =$$
$$= r + tt'r' \exp(i\delta) \lim_{m\to\infty} \frac{1-(r')^{2(m-1)} \exp(i(m-1)\delta)}{1-(r')^2 \exp(i\delta)} \qquad (3.13)$$

By using elementary math, (3.13) can be rewritten in the following form:

$$A^{(r)} = -\frac{r'\{1-[(r')^2 + tt']\exp(i\delta)\}}{1-(r')^2 \exp(i\delta)} \qquad (3.14)$$

In this case the reflectance $R_{FP} = |A^{(r)}|^2$ of such an Fabry-Perot interferometer will be

$$R_{FP} = \frac{4R\sin^2\left(\frac{\delta}{2}\right)}{(1-R)^2 + 4R\sin^2\left(\frac{\delta}{2}\right)} \qquad (3.15)$$

Similarly, the amplitude of wave $A^{(t)}$ transmitted through Fabry-Perot interferometer will be the sum of all waves escaped into the second medium and, according to (3.10)

$$A^{(t)} = tt' \lim_{m\to\infty} \sum_{k=1}^{m} (r')^{2(k-1)} \exp(i(k-1)\delta) =$$

$$= tt' \lim_{m\to\infty} \frac{1-(r')^{2m}\exp(im\delta)}{1-(r')^2\exp(i\delta)} = \frac{tt'}{1-(r')^2\exp(i\delta)} \qquad (3.16)$$

The transmittance of the Fabry-Perot interferometer will be $T_{\mathrm{FP}} = |A^{(t)}|^2$:

$$T_{\mathrm{FP}} = \frac{T^2}{(1-R)^2 + 4R\sin^2\left(\frac{\delta}{2}\right)} \qquad (3.17)$$

The transmission spectra (numerically calculated according to (3.17)) of a 6.5 μm length air-filled Fabry-Perot cavity for different mirror reflectivities at normal incidence is given in figure 3.8. The transmittance of a Fabry-Perot interferometer reaches maximum at such values of δ when the following condition is satisfied:

$$\delta/2 = (2\pi/\lambda)\, n_1 d \cos\theta_1 = m\pi, \qquad m = 0, \pm 1, \pm 2, \ldots \qquad (3.18)$$

and reaches minimum at the following values of δ:

$$\delta/2 = (2\pi/\lambda)\, n_1 d \cos\theta_1 = m\pi/2, \qquad m = \pm 1, \pm 2, \ldots \qquad (3.19)$$

The peaks of transmittance (3.18) are known as *fringes* and m is called the order of corresponding fringe. It can be proven that the maximum values of transmittance at the peaks (3.18) reach unity if the Fabry-Perot interferometer is constructed from nonassertive materials.

As illustrated in figure 3.8, the ratio of the width of the fringes to interfringe spacing decreases with the increase of reflectivity of the mirrors.

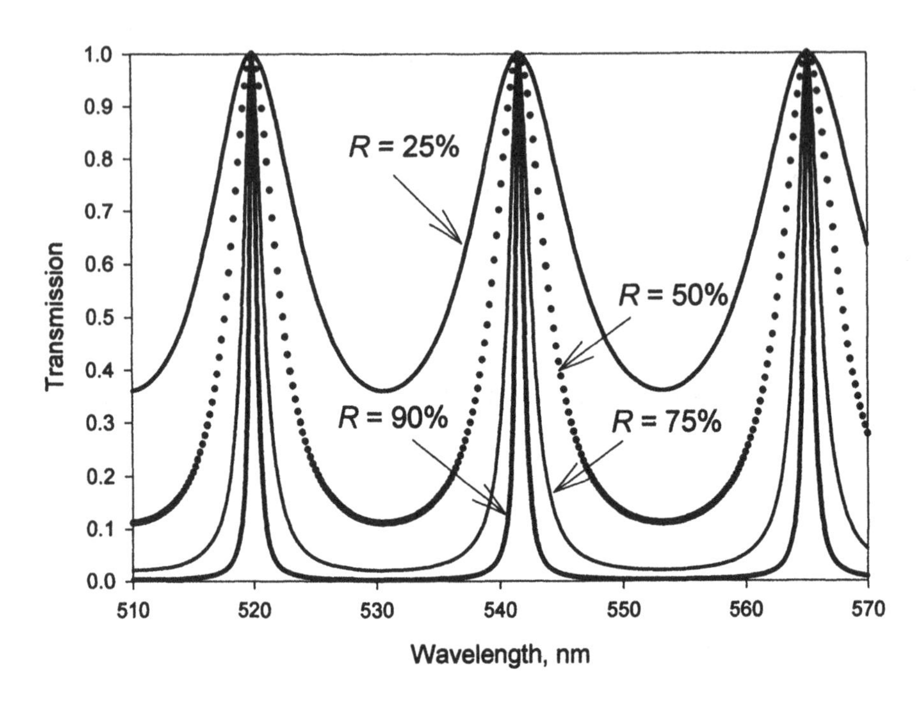

Figure 3.8. Transmission spectra of 6.5 μm length air-filled FP cavity for different mirror reflectivities.

The quantitative parameter to characterize this ratio is the *finesse* F_{FP} of Fabry-Perot interferometer:

$$F_{\mathrm{FP}} = \frac{4R}{(1-R)^2} \tag{3.20}$$

For the air-filled Fabry-Perot cavity, the transmittance T_{FP} can be expressed through the finesse F_{FP} (see (3.17) and (3.20)):

$$T_{\mathrm{FP}} = \frac{1}{1 + F_{FP}\sin^2\frac{\delta}{2}} \tag{3.21}$$

If we define the width of the fringe (the distance between the points at the fringe where $T_{\mathrm{FP}} = 0.5$) as Δ (i.e., $T_{\mathrm{FP}} = 0.5$ at $\delta = 2m\,\pi \pm \Delta/2$), then at high finesse of Fabry-Perot interferometer Δ can be expressed through F_{FP}:

$$\Delta = \frac{4}{\sqrt{F_{FP}}} \tag{3.22}$$

The classical application of Fabry-Perot interferometers is the examination of the fine structure of spectral lines. The fringes are produced by passing light from the evaluated source through the interferometer. Measurement of the fringe pattern as a function of the physical parameters of the etalon yields very precise values of the wavelengths of the various components of the line [18]. Also, Fabry-Perot interferometers serve as laser resonators if they are filled at least partially by an optically active medium (see, for example, [36]). In addition to these applications, new use of Fabry-Perot etalons as an optical filter in the feed-back loop for the stabilization of the tunable lasers emerges.

3.3.2 Multilayer Dielectric Reflectors

The optical principle that serves as the base for multilayer dielectric reflectors is the same multiple beam interference that was investigated in the previous section for the Fabry-Perot interferometer case.

The classic design of high-reflectance coating is based on alternating quarter-wave layers of two different materials. Examples of such stacks have been already discussed in sections 2.5.3 and 2.9 of this book. The high reflectance in quarter-wave stack takes place because the beams, reflected from all the interfaces in the multilayer, are in phase when they reach the front surface where the constrictive interference of all the reflected waves occurs. The peak reflectivity of such a multilayer stack was given in section 2.5.3 for the large number of layers in the stack. For any number of layers the reflectivity is equal to [21]

$$R = \left(\frac{1 - \frac{n_s}{n_0}\left(\frac{n_2}{n_1}\right)^{2N}}{1 + \frac{n_s}{n_0}\left(\frac{n_2}{n_1}\right)^{2N}} \right)^2 \tag{3.23}$$

where n_1 and n_2 are the refractive indices of the alternative layers, n_0 is the refractive index of incident medium; n_s is the refractive index of the substrate, and N is the number of pairs of alternating layers in the stack. The reflectance according to (3.23) quickly approaches unity as the N grows (see figure 2.7 of section 2.5.3).

The reflectance spectrum of a 17-layer stack of titanium dioxide (n = 2.3) and magnesium fluoride (n = 1.38) on glass substrate at normal incidence is presented in figure 3.9

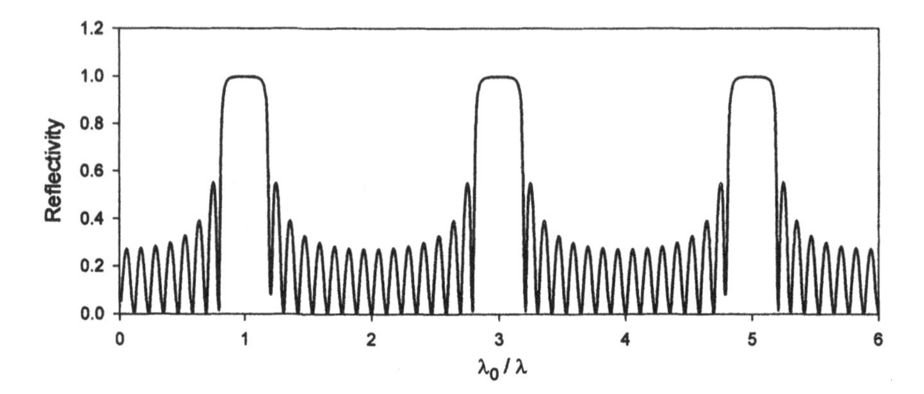

Figure 3.9 Reflectance of a 17-layer stack of titanium dioxide (n = 2.3) and magnesium fluoride (n = 1.38) on glass at normal incidence.

As with the Fabry-Perot etalon (figure 3.8), the reflectance spectrum of such a multilayer contains multiple reflection peaks corresponding to the phase-matching conditions of the reflected waves at the different interfaces. However, these reflection maximums are considerably wider than that of the Fabry-Perot etalon. The width of the high reflectance plateau depends on the refractive index contrast between low and high refractive index materials, composing the high reflectance multilayer. It can be shown (see, for example, [34]) that the width of the first (or fundamental) high reflectance zone (see figure 3.9) for dielectric multilayer will be

$$\frac{\Delta\lambda}{\lambda} = \frac{4}{\pi}\sin^{-1}\frac{n_H - n_L}{n_H + n_L} \tag{3.24}$$

where n_H and n_L are the refractive indices of high and low refractive index materials composing the high reflectance multilayer. In figure 3.10 the numerically calculated (according to (3.24)) bandwidth of the fundamental high-reflectance zone of quarter-wave stack is given as a function of the refractive index ratio of the layers composing the stack.

For the visible region of spectrum, the common materials composing the high-reflectance multilayer are zinc sulphide (n = 2.35), cryolite (n = 1.35), titanium dioxide (n = 2.3), magnesium fluoride (n = 1.38), and silicon dioxide (n = 1.5 at visible spectral range). For the infrared high refractive index, semiconductors became available as the wavelength exceeds the absorption band edge of semiconductors.

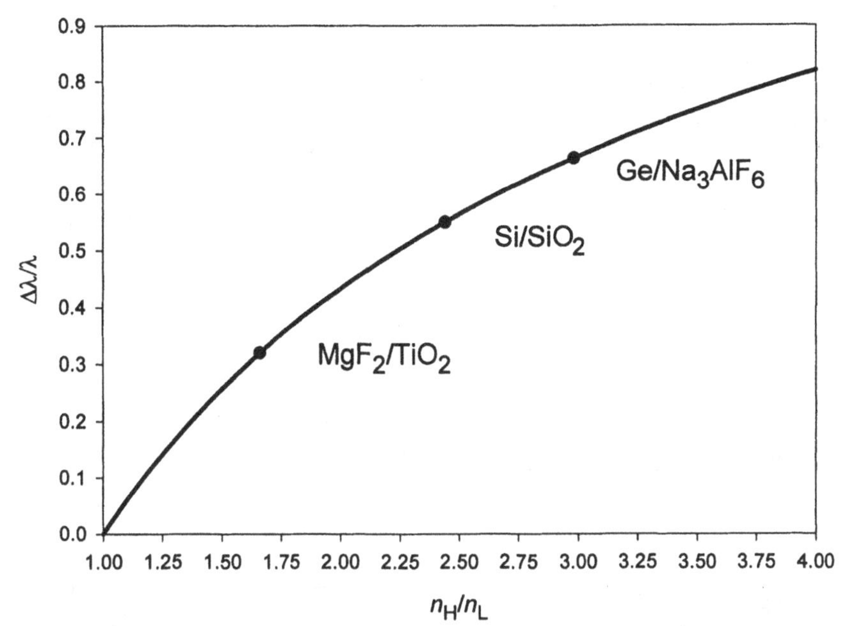

Figure 3.10. The bandwidth of the fundamental high-reflectance zone of quarter-wave stack as a function of the refractive index ratio of the layers composing the stack.

Silicon (n = 3.5) can be used starting from approximately 1.1 μm, while germanium (n = 4) can be used for the spectral region of 1.8 to 3 μm.

As follows from the above discussion, the high-reflectance dielectric multilayer can be constructed to have very high reflectance over a wide range of wavelengths. However, such multilayers have some disadvantages. One of the disadvantages is that the high reflectance zone of such a reflector, although it can be made wide, is still limited (see figure 3.9). Moreover, since the reflectance peaks are located where the waves, reflected from each interface in multilayer, are in phase, the wavelength positions for said peak should depend on the angle of incidence, similar to the Fabry-Perot etalon case. Figure 3.11 presents the numerically calculated reflectance spectra of a 17-layer stack of titanium dioxide and magnesium fluoride on glass for different angles of incidence. Both angular and polarization influences of reflection spectra are illustrated. The wavelength position of the reflection peak can be found from (3.18):

$$\lambda = \lambda_0 \cos \theta' \qquad (3.25)$$

where λ_0 is the peak reflection wavelength at the normal incidence and θ' is the angle in the medium. While for the Fabry-Perot etalon case, $\cos \theta'$ can be easily determined according to the formula $\sqrt{1-\left(\sin\theta/n\right)^2}$, where n is the refractive index of the medium between the mirrors (see figure 3.7), for the high-reflectance multilayer it is not that simple, since it consists of alternating layers of high and low refractive materials. It is usually accepted that for this case n in (3.25) is the effective refractive index of the multilayer that is equal to $(n_H + n_L)/2$.

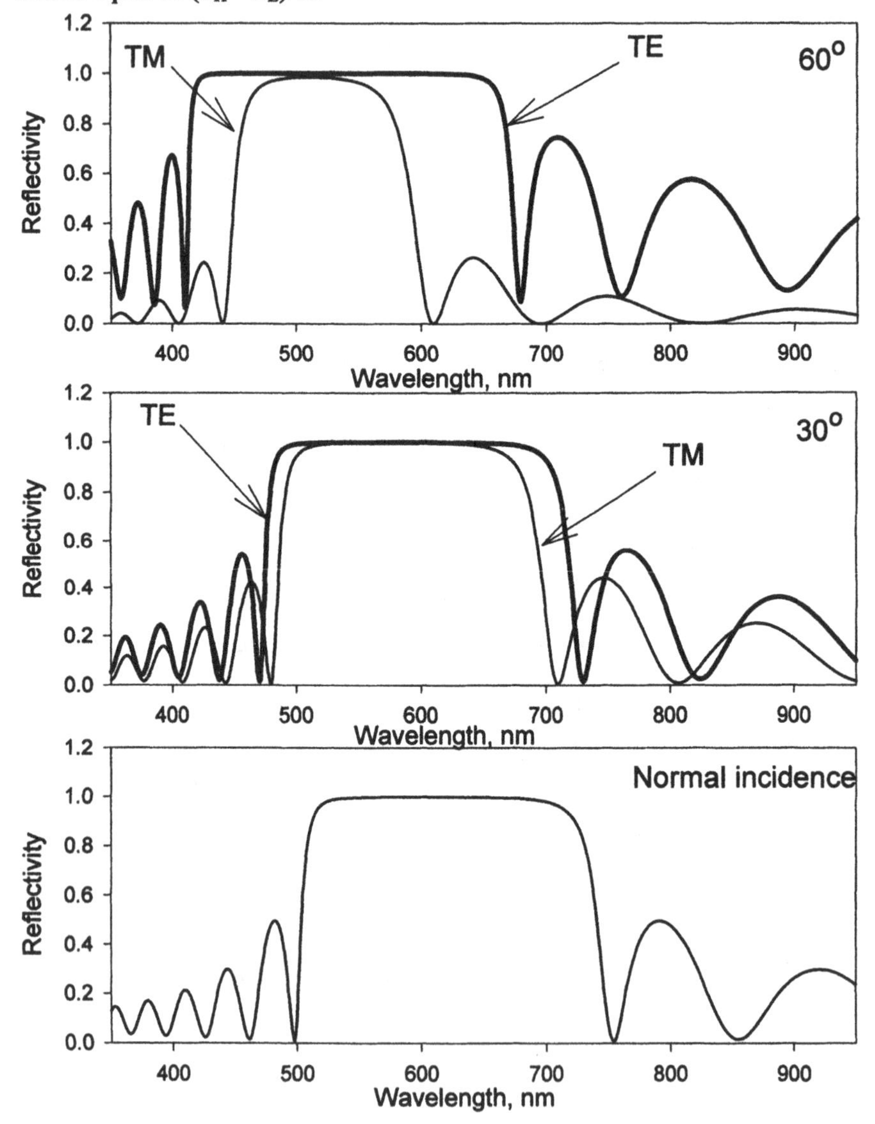

Figure 3.11. Reflectance of a 17-layer stack of titanium dioxide (n = 2.3) and magnesium fluoride (n = 1.38) on glass at different angles of incidence.

The quarter-wave stack cannot be used in some applications. Several methods for extending the high reflectance zone (we will denote these methods as "classical," since a new, "nonclassical" methods for solving this problem will be discussed in chapter 4 of this book) have been created.

One of these methods involves changing the thickness of successive layers throughout the multilayer to form a regular progression to ensure that at any wavelength within a reasonably wide range, there are enough layers in the multilayer that have optical thickness sufficiently near a quarter-wave to give high reflectance. Such an approach can provide reasonably wide high-reflectance ranges (a 35-layer filter of zinc sulphide and magnesium fluoride could provide a reflection band from 300 to 826 nm [37]). However, it requires a large number of layers in multilayers, and the overall reflectance is usually around 95% over that range. Detailed discussion of such an approach with good examples can be found in [34].

Another method involves placement of a quarter-wave stack for one wavelength on top of another for a different wavelength ([38]). To avoid Fabry-Perot-like peaks in the resultant reflectance spectrum, it was found that the low index layer, one quarter-wave thick at a mean wavelength (i.e. wavelength equally spaced to the wavelengths of both stacks) should be placed in between stacks. That is, the wide-band high-reflectance multilayer should have the form
{Air | $(1 - x)$ $(H\,L\,H \ldots H\,L\,H)$ L $(1 + x)$ $(H\,L\,H \ldots H\,L\,H)$| Substrate}, where L and H denounce the low and high refractive index layers and quarter-wave on that mean wavelength.

The numerically calculated reflectance of such a multilayer reflector is presented in figure 3.12. The structure of this multilayer was as follows. The silicon dioxide/silicon multilayer on glass substrate had the following structure: {Air | 0.8 $(H\,L\,H\,L\,H\,L\,H\,L\,H)$ L 1.2 $(H\,L\,H\,L\,H\,L\,H\,L\,H)$| Glass}, wherein the low refractive index material was silicon dioxide, while high-refractive index material was silicon. The mean wavelength was 1600 nm. It should be noted that in this example the wavelength-dependent absorption of silicon was not taken into account. Hence, the reflectance spectra below ~1100 nm will be different, since silicon is transparent only from ~1050 to 1100 nm. It is clearly illustrated that using such a method near 100% reflectance can be obtained over a wide wavelength range. In addition, at the mean wavelength of such a dielectric reflector, high reflectance takes place for all angles of incidence. Such reflectors are suitable for most of the present applications. However, if one would want to extend the high-reflectance range further (or to extend the wavelength range, where high reflectance takes place for both polarizations for all possible angles of incidence), the number of layers needed increases dramatically, and a different approach is needed. Such an approach is considered in chapter 4 of this book.

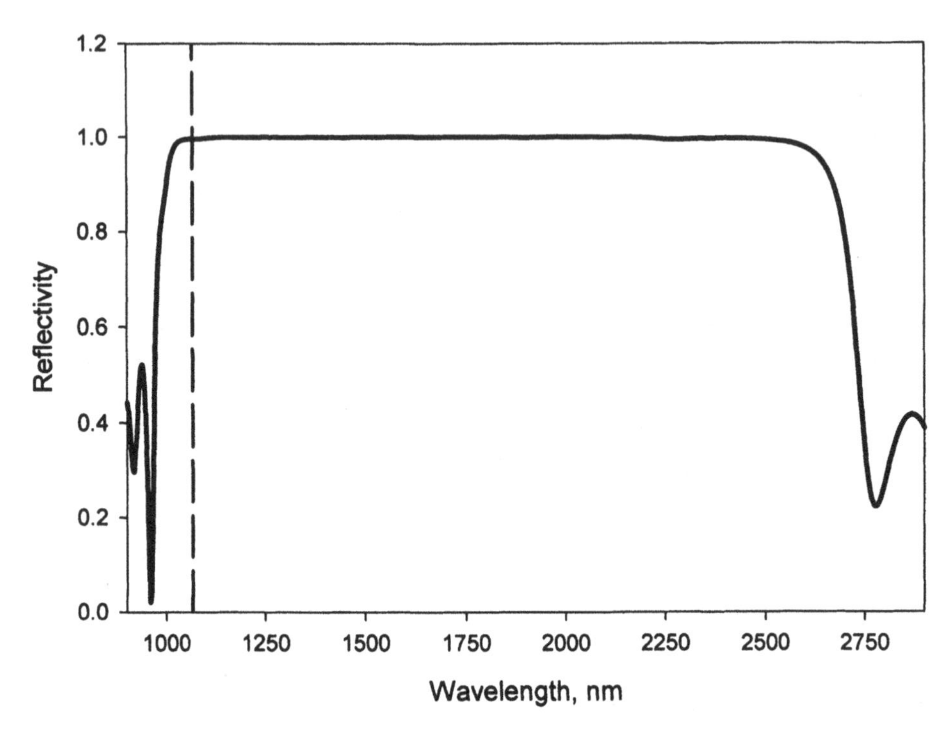

Figure 3.12. Numerically calculated reflectance of the silicon dioxide/silicon multilayer on a glass substrate having following structure: {Air | 0.8 (*H L H L H L H L H*) *L* 1.2 (*H L H L H L H L H*) | Glass}.

3.4 Edge Filters

Edge filters are filters that provide an abrupt change between a region of rejection and a region of transmission. They can be divided into two main groups: long-pass and short-pass filters. *Long-pass filters* are edge filters that transmit light at the wavelengths greater than some characteristic wavelength, known as an *edge wavelength*, while they reject (reflect or absorb) the light at the wavelengths below the edge wavelength. *Short-pass filters*, in reverse, transmit the light having the wavelengths below the edge wavelength, while they rejecting light above the edge wavelength. There are two main types of edge filters according to their principle of operation: *absorption-based filters* (i.e., filters where the rejection of light is caused by absorption in filter material) and *interference-based filters* (i.e., filters where the rejection of light is caused by reflectance from multiple layers composing the filter).

I do not attempt to give a detailed review of different edge filter designs and types (since a lot of them were developed) but rather limit myself to a very brief description of the basics to give the reader minimal understanding of this topic. For more information readers should see, for example [34].

3.4.1 Absorption Filters

Absorption filters consist of a thin film or slide of material that has an absorption edge at the required wavelength. The typical example of a long-pass absorption filter is the filter that utilizes semiconductor material. Semiconductors are known to have an absorption band that extends to some characteristic wavelength, which corresponds to the bandgap energy of a particular semiconductor. As an example, the absorption coefficient of silicon is presented in figure 3.13. Note the logarithmic scale. In figure 3.14 the numerically calculated transmission spectra of silicon thin film for different film thicknesses is presented. The reflection at both surfaces hasn't been taken into account to enhance the absorption effect.

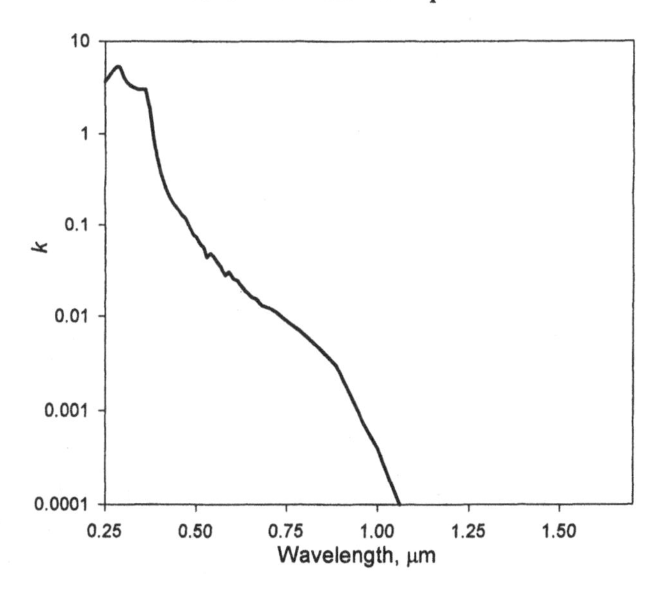

Figure 3.13. The imaginary part of a silicon refractive index (absorption coefficient) (data taken from [39]).

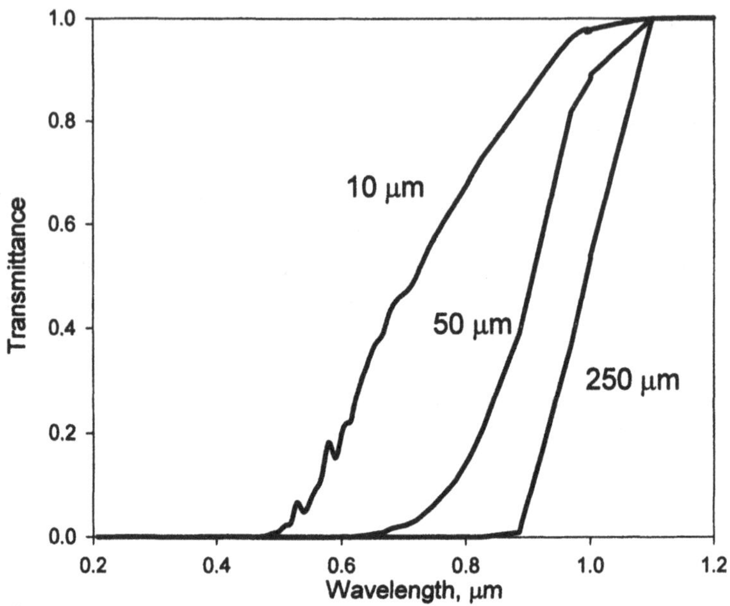

Figure 3.14. Numerically calculated transmission spectra of silicon thin film for different film thickness (data on the Si absorption coefficient taken from [39]).

One can see that the transmission/rejection edge could be made very sharp ($\Delta\lambda/\lambda \sim 10\%$) for a silicon layer thickness above 250 μm. Similar results can be obtained with other semiconductors. Since the absorption band edge of different semiconductors and semiconductor composites can vary from ~ 500 nm for gallium phosphate (GaP) and aluminum arsenate (AlAs) to more than 2 microns for indium arsenate (InAs) and InSb, and, more important, the absorption band edge can be smoothly tuned by adjusting the semiconductor composition (for example, $Al_xGa_{1-x}As$ absorption band edge tunes quite linearly from 2.1 eV for x =1 to 1.4 for x = 0), long-pass filters can be obtained with a reasonably sharp edge for any wavelength at the ~500 to 2400 nm range. However, semiconductor-based long-pass filters have a significant disadvantage – high reflection losses (see figure 2.2) caused by the high refractive index of semiconductor materials. If the reflectance at both surfaces of the silicon wafer is taken into account, then the maximum transmittance of the filters in figure 3.14 will not exceed 50%. Such a problem is usually solved by antireflection coatings of both semiconductor surfaces in the case of semiconductor wafer (or slide) used (see section 3.1 of this book) or, in the case of thin film semiconductor material, of the top and between the substrate and the layer of semiconductor material. As was shown in section 3.1, using such an approach the reflection losses can be suppressed to sufficiently low values. However, such an approach also has some significant disadvantages: while the absorption edge shape and position of uncoated semiconductor film does not depend on angle of incidence (only absorption value changes), both the absorption edge shape and position of antireflection-coated semiconductor film does depend on angle of incidence (see the discussion at the end of section 3.1) that is, can be used effectively only for some limited angular range. In addition, semiconductor absorption edges show strong temperature dependence (in fact, this effect is the base of some fiber-optic temperature sensors).

Other materials that are used to form absorption filters include colored-glass filters (for example, Schott glass filter), which operate through the process of either ionic absorption of inorganic material, dispersed uniformly through the glass slide or through the absorptive scattering of crystallites formed within the glass. Such filters offer quite wide design freedom in terms of absorption edge position and can be arranged either in short-pass or long-pass forms (contrary to semiconductor based filters, which are essentially long-pass filters). The considerably lower refractive index of such filters than that of semiconductor filters make the reflection losses lower. Temperature dependence of the rejection band edge position of such filters is also lower than that of semiconductors. However, such filters also suffer from some limitations: the transmission of such filters rarely extends below 300 nm, which make them not suitable for deep UV applications, and the transmission through the transparency range of such filters is usually not very uniform.

The common disadvantage of all absorption-based edge filters is their instability under high-power (and even medium-power) radiation at the wavelengths of absorption band. For most laser applications such filters are not suitable. For such applications a different type of edge filters is usually used – interference edge filters.

3.4.2 Interference Edge Filters

Quarter-wave stack (see section 3.3.2) can be considered as the basic type of interference edge filters. The transmission spectrum of the quarter-wave stack contains alternative low- and high-reflectance zones and, hence, alternative high- and low-transmittance zones. As an example, the numerically calculated transmittance spectrum through quarter wave at 1100 nm stack consisting of 21 layer cryolite (n = 1.35) / titanium dioxide (n = 2.35) on glass substrate is presented in figure 3.15 (dashed line).

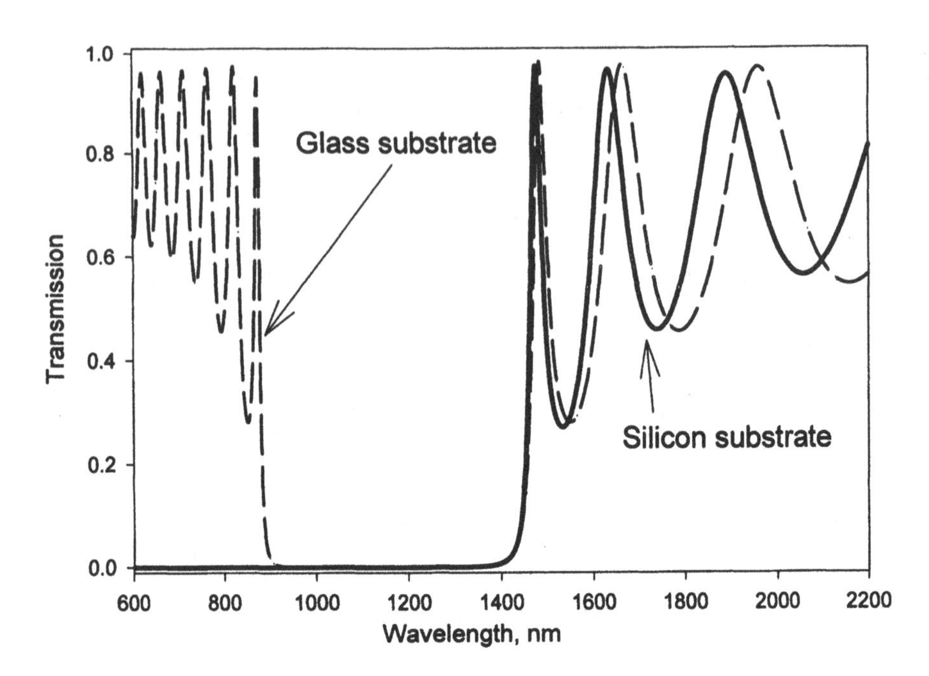

Figure 3.15. Numerically calculated transmission spectra of 21 layer quarter-wave stack at 1100 nm of cryolite (n = 1.35) / titanium dioxide (n = 2.35) on silicon (solid line) and glass (dashed line) substrates (data on Si absorption coefficient taken from [39]).

Such a filter can be used as a long-pass filter with an edge at ~ 1400 nm or as a short-pass filter with an edge at 900 nm. These wavelengths can be tuned by changing the wavelength, at which the stack is quarter-wave. Such an edge filter will be suitable for relatively narrow-band applications that is, when the width of rejection zone is higher than the spectral width of light to be eliminated. For all other cases, required elimination of all wavelengths shorter than (or longer than) a particular value different filter design is needed. Such an edge filter can be constructed by coupling the interference filter with the absorption filter of the previous section.

The example of combined interference/absorption edge filter can be imagined as the quarter-wave stack on the silicon substrate discussed above. The numerically calculated transmittance spectrum through quarter wave at 1100 nm stack consisting of 21 layer cryolite (n = 1.35) / titanium dioxide (n = 2.35) on silicon substrate is presented in figure 3.15 (solid line). Such a combined filter will provide truly long-pass characteristic spectrum with the edge at ~ 1400 nm.

By using combined interference/absorption filter design, one can obtain the filters that have deep rejection of absorption filter and flexibility in edge position and sharpness of interference filter. However, it should be noted that combined filter will also combine the disadvantages of both filters: it will not be suitable for high-power illumination due to absorptance (as of absorption filter) and will have significant dependence of the edge position on the angle of incidence (as of interference filter).

The important interference edge filter property that is clearly illustrated by Figure 3.15 is the presence of so-called *ripple*. Under the ripple it is usually underhanded the strong variation of transmittance with the wavelength near the edge of such a filter. Ripple is caused by the multiple wave interference that give us the interference filter rejection zone. Filters with such a strong ripple that shown in figure 3.15 are obviously limited in applications. To remove ripple, the structure of high-reflectance multilayer stack is usually modified.

To understand the methods developed to suppress the ripple in interference edge filters, one need to understand the equivalent optical index theory, which is beyond the scope of present book. Readers can find it in, for example, [34] or [40]. Here I present just a very brief illustrative example showing the result rather than the method.

The simplest way to suppress the ripple is to add a pair of eighth-wave layers to the stack, one between the air and a stack, another between the substrate and the stack. The numerically calculated transmittance spectrum of such a stack having structure {Air | *H*/2 *L H L H L H L H L H L H L H L H L H L H L* *H*/2| Glass} is presented in figure 3.16. The high-index material was assumed to be titanium dioxide, while the low index material was assumed to be glass and the stack was assumed to be quarter-wave at 1 μm. If compared with figure 3.15, considerable suppression of the ripple at long-wavelength edge is clearly illustrated. The indices of materials composing multilayer should be chosen according to the material of the substrate and should be different for different substrates. However, available optical materials offer rather discrete values of refractive indexes, so the thickness of the layers composing the multilayer stack is usually altered. Several methods of optimization of interference edge filter structure have been developed. Probably the most common of them is the computer refinement method [40]. It consists of numerical multiparameter optimization, wherein the parameters to optimize are layer thicknesses, and it usually gives good results.

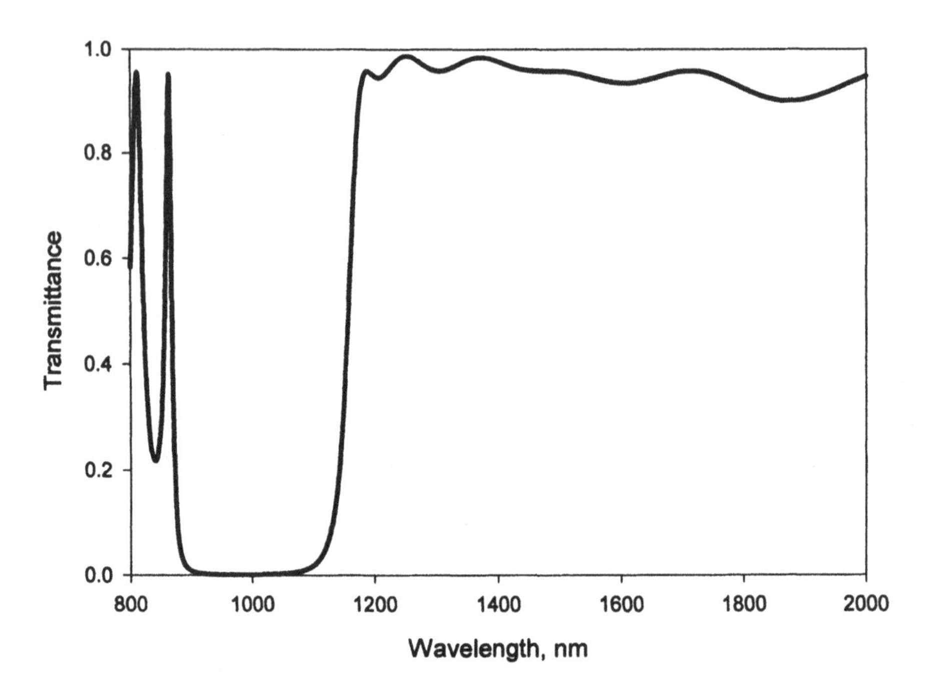

Figure 3.16. A numerically calculated transmittance spectrum of an interference long-pass filter with suppressed ripple.

As in multilayer dielectric reflectors, discussed in section 3.3.2 of this book, practical interference edge filters require an extended rejection zone for some applications and an extended transmission zone for other applications.

The ways to extend the rejection zone of interference edge filters are the same in their nature as the methods used to extend the high-reflectance zone of multilayer dielectric reflectors. To do that, one needs to place several multilayer stacks in series. As in section 3.3.2, the intermediate layers or other methods can be used to optimize the rejection uniformity. However, in the case of interference edge filters this optimization is more complex than in multilayer dielectric reflectors, since not only should Fabry-Perot peaks be avoided within the rejection zone, but also the ripple should be suppressed near the transmittance edge.

Extending of the transmission zone is a more complex procedure. It involves the suppression of several reflection orders of multilayer stack (see figure 3.9). To do that complex multilayer structures were suggested instead of multilayers contain alternative layers of just two materials [41]. However, such filters are quite hard to produce due to the high number of layers needed.

Another important parameter of interference edge filters is edge steepness. The edge steepness is proportional to the number of pairs of alternative layers used in multilayer stack, and for the same number of layers it is proportional to the refractive index contrast between low- and high-refractive index materials, composing multilayer. The numerically calculated edge steepness ($dT/d\lambda$ [nm^{-1}]) of the filters of the same structure shown in figure 3.16 is presented in figure 3.17 as a function of a number of layers in multilayer stack for silicon/silicon dioxide and silicon dioxide/titanium dioxide stacks. One can see that edge steepness experienced exponential growth with the number of pairs used. In fact, for most applications, the number of layers necessary to produce the required rejection in the rejection band of the filter is sufficient to produce the edge steepness needed. However, if higher edge steepness is required, the easiest way of improving it is to increase the number of layers in multilayer stack. Using such an approach, however, will bring some disadvantages in interference edge filter performance. The ripple in general will be enhanced, and the wavelength position of the edge will shifted toward longer wavelengths for long-pass filters and toward shorter wavelengths for short-pass filters.

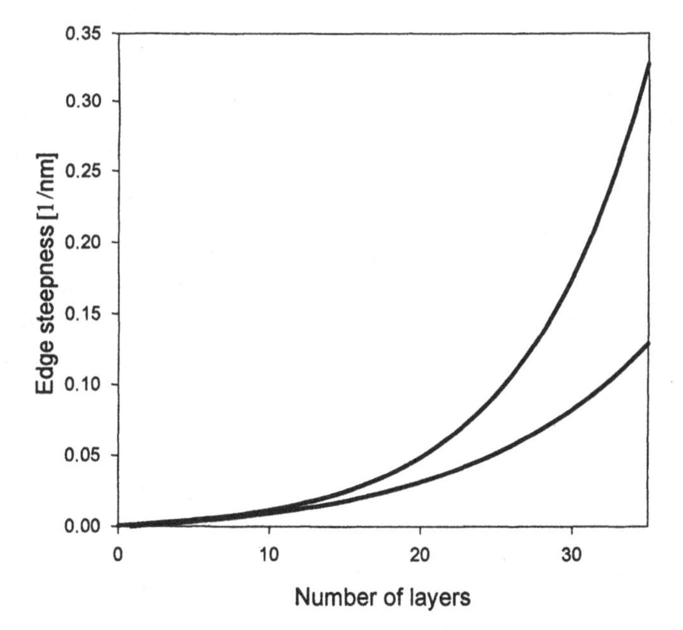

Figure 3.17. Numerically calculated edge steepness of the filter of figure 3.16 as a function of a number of layers in multilayer stack for silicon/silicon dioxide and silicon dioxide/titanium dioxide stacks.

The alternative approach to increasing interference edge filters edge steepness is to use multilayer stacks, working in higher reflection orders (see figure 3.9). However, such an approach also has some disadvantages. The width of the rejection zone is generally decreased with the number of reflection order that is, higher number of layer pairs needs to be used to get the same level of rejection zone width. This is disadvantageous from both the cost and manufacturability points of view. The most important disadvantage is that the acceptable errors in layer thickness are reduced with the multilayer reflection order number. According to Macleod [34], the acceptable random errors in layer thickness for first reflection order multilayer stack is within the range of 5 to 10% for most of filters used, while, for example, for the fifth reflection order filter this error is within the range of 1 to 2%.

3.5 Band-Pass Filters

Band-pass filters are filters that provide a wavelength-limited region of transmission surrounded on each side by rejection regions. Band-pass filters can be divided according to the width of transmission zone into *broadband-pass filters* and *narrowband-pass filters*. There is no definite boundary between these two types of band-pass filters. In this book we follow Macleod [34] and under broadband-pass filter assume such a band-pass filter, having the bandwidth of 20% or more of transmission band central wavelength and constructed by combining long-pass and short-pass filters.

The best arrangement of broadband-pass filter is to place two multilayer stacks (short-pass and long-pass) on the opposite sides of a single substrate.

To give maximum transmission within the pass-range, each multilayer should be designed to match the substrate. As an example, in figure 3.18b the numerically calculated transmittance spectra of the broadband-pass filter is composed of two multilayer stacks (long-pass and short-pass) made of magnesium fluoride/titanium dioxide on the glass substrate, while in figure 3.18a the transmittance spectra of individual stacks composing the filter of figure 3.18b are given.

Broadband-pass filter can also be arranged by depositing both long-pass and short-pass multilayer stacks on the same side of the substrate. The design problem that arises in this case is the appearance of transmission deeps in the pass zone and transmission peaks in the rejection zone due to interference between waves reflected from long-pass and short-pass stacks. This problem is usually solved by modification of the long-pass and short-pass stacks structure. In addition, in such broadband-pass filter design special care should be taken to maximize the transmittance within the filter pass zone. In general, the transmittance through such a filter will not exceed the value

$$T_{\max} = \frac{T_{sp} T_{lp}}{\left[1 - \sqrt{R_{sp} R_{lp}}\right]^2} \tag{3.26}$$

where T_{sp} and R_{sp} are transmittance and reflectivity of short-pass multilayer stack and T_{lp} and R_{lp} are transmittance and reflectivity of short-pass multilayer stack.

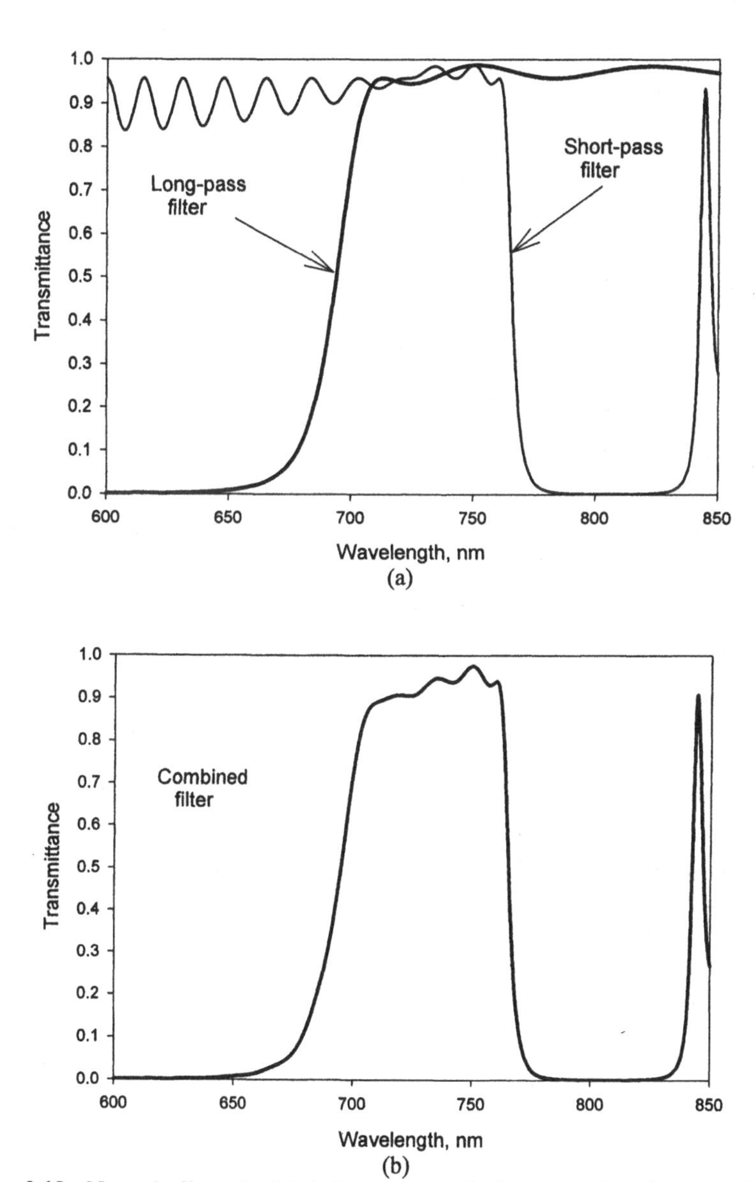

Figure 3.18. Numerically calculated long-pass and short-pass interference edge filters transmittance spectra (a) and the transmittance spectrum of the broadband-pass filter, composed from edge filters (b).

3.6 Narrowband-Pass Filters

The simplest design of a narrowband-pass filter is the Fabry-Perot filter (see figure 3.8), discussed in section 3.3.1. However, the pass-band shape of the Fabry-Perot filter is triangular. In addition, in its original design it requires two precisely aligned and spaced high-flatness plates, which is not practical for many applications. Therefore, the narrowband-pass filters are usually made in slightly modified form with respect to Fabry-Perot etalon: Fabry-Perot thin film filters. A *Fabry-Perot thin film filter* is a thin film assembly consisting of a dielectric layer bounded by either two metallic reflecting layers or by two multilayer dielectric reflectors of section 3.3.2. Realization of a Fabry-Perot filter with metallic reflective layers is called *metal-dielectric Fabry-Perot filter*, while a Fabry-Perot filter with two dielectric multilayer reflectors is called an *all-dielectric Fabry-Perot filter*.

3.6.1 The Metal-Dielectric Fabry-Perot Filter

The metal-dielectric Fabry-Perot filter is the simplest realization of narrowband-pass filters. In such a filter, the dielectric layer, surrounded by metal reflection layers, serves as a spacer in Fabry-Perot etalon and therefore, is called *spacer layer*. The metallic reflection layers have to provide reasonably high reflections at the surface while keeping losses as low as possible. For the visible region of the spectrum silver is an optimal metal (see figure 3.4), while for ultraviolet and deep ultraviolet aluminum is the preferred material. The theory for such filters is exactly the same as it was presented for Fabry-Perot etalon in section 3.3.1, except for the presence of absorption. I do not repeat the theory here and focus on the absorption problem only.

Absorption of light is the biggest disadvantage of metal-dielectric narrowband-pass filters. Although absorptance during the single reflection from thin silver film is small and absorptance during transmission also can be minimized by using thin metal films, in Fabry-Perot cavity absorptance is greatly enhanced due to multiple reflections of the transmitted light. As an illustration, numerically calculated metal-dielectric Fabry-Perot filter transmittance (a) and absorptance (b) spectra are presented in figure 3.19 for different thicknesses of metal (silver) layers.

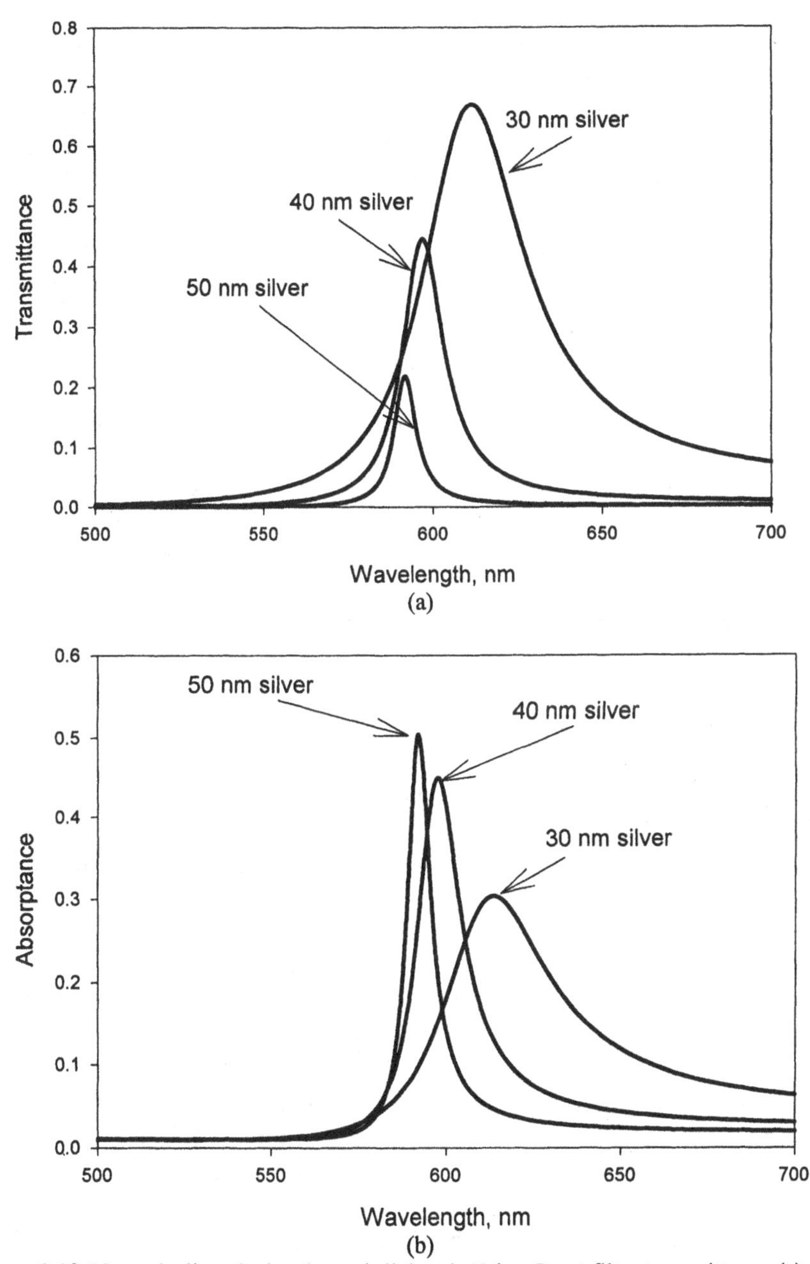

Figure 3.19. Numerically calculated metal-dielectric Fabry-Perot filter transmittance (a) and absorptance (b) spectra for different thicknesses of metal (silver) layers.

Figure 3.19 clearly illustrates the tradeoff that always exists for metal-dielectric Fabry-Perot filters design: to get a really narrow transmission band one needs to keep Fabry-Perot cavity finesse high (see (3.22)), while the absorptance in Fabry-Perot cavity increases with the finesse so the peak achievable transmittance decreases. According to figure 3.19, for the first-order Fabry-Perot cavity the maximum achievable transmittance of the filter, designed to work at ~600 nm with the transmission bandwidth of ~10 nm, will not exceed 25% (in practice this number will be even lower). However, the situation is not as bad for higher-order Fabry-Perot filters. The bandwidth of higher-order Fabry-Perot transmission peaks decreases faster than absorption increases with the number of order. According to [42], for the magnesium fluoride/silver metal-dielectric filter the transmission peak with the bandwidth of 7 nm, maximum transmittance of 26% was experimentally obtained. In practice, up to the third-order Fabry-Perot filters are usually used. To suppress lower-order transmittance peaks such filters can be used with the short-pass edge filter.

Metal-dielectric Fabry-Perot filters have the same dependence of the transmittance peaks wavelength as the Fabry-Perot cavity (see formula (3.25)). In addition to this disadvantage, the absorptance of such a filter strongly depends on angle of incidence. If adding quite poor overall transmittance, such filters become quite poor performers. They are usually used in such applications where other filters (such as all-dielectric Fabry-Perot filters) are prohibited by either cost or other factors (such as in deep UV).

3.6.2 The All-Dielectric Fabry-Perot Filter

As was mentioned above, in the all-dielectric Fabry-Perot filter the metallic reflecting layers are replaced by high-reflecting dielectric multilayers. The structure of the all-dielectric Fabry-Perot filter is schematically shown in figure 3.20.

As follows from figure 3.20, two different cases of such a filter should be considered: {Air| *H L H L* ... *H L* ***H H*** *L H* .... *L H L H*| Substrate} and {Air| *H L H L* ... *L H* ***L L*** *H L* .... *L H L H*| Substrate}(the refractive indices of layers adjacent to air and substrate should be high to maximize the reflection from multilayer). The transmission spectrum of all-dielectric Fabry–Perot will be a narrow maximum within the broad minimum. As an example, in figure 3.21 the numerically calculated transmittance spectrum of a 17-layer titanium dioxide/magnesium fluoride narrowband-pass Fabry-Perot filter on glass substrate is presented for the high refractive index (titanium dioxide) spacer layer.

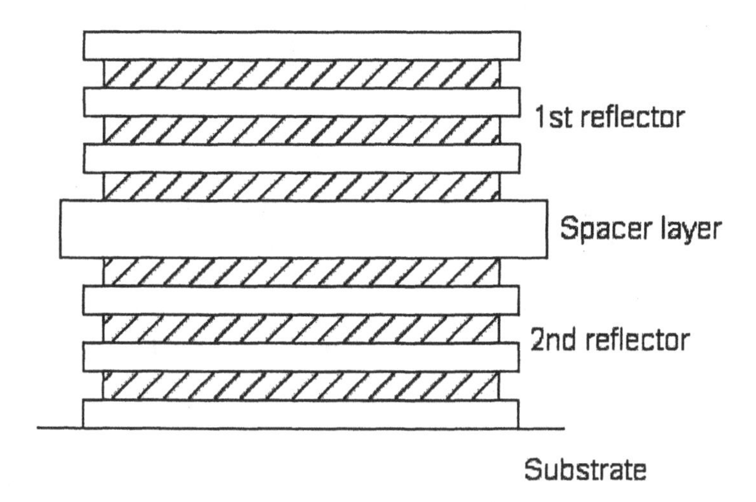

Figure 3.20. The structure of an all-dielectric Fabry-Perot filter.

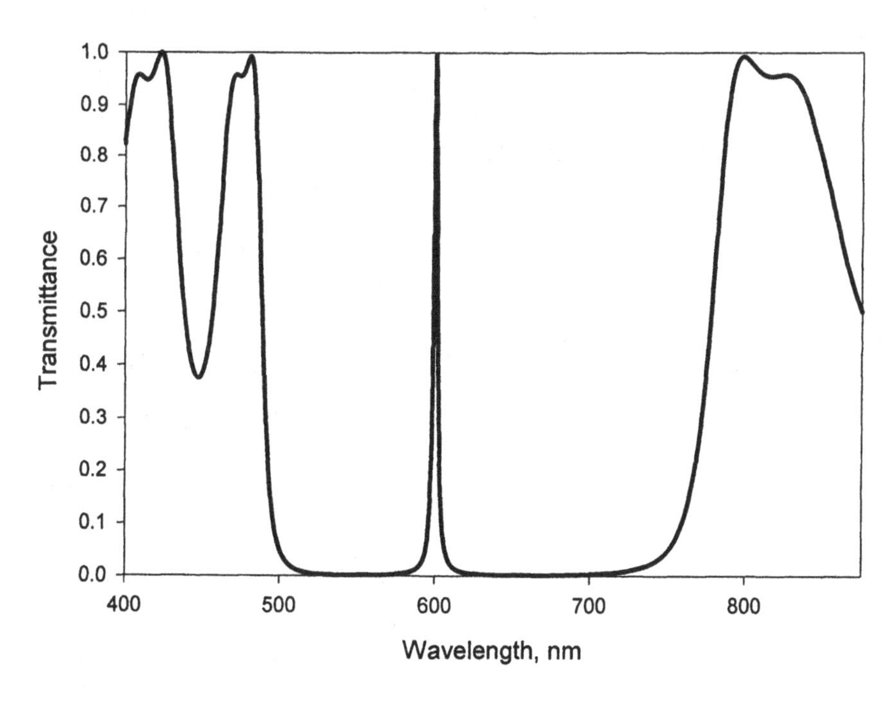

Figure 3.21. Numerically calculated transmittance spectrum of 17-layer titanium dioxide/magnesium fluoride narrowband-pass Fabry-Perot filter on glass substrate.

The width of the maximum and the transmittance at the maximum will depend on the reflectivities of two multilayer stacks. It can be mathematically derived (see, for example [34]) that the bandwidth of all-dielectric Fabry-Perot filter for high refractive index spacer layer will be

$$\frac{\Delta\lambda_H}{\lambda_0} = \frac{4n_L^{2x} n_s}{m\pi n_H^{2x+1}} \qquad (3.27)$$

where n_L and n_H are refractive indices of low- and high-refractive index layers in multilayer stacks; n_s is the refractive index of spacer layer; λ_0 is the wavelength of the transmittance maximum; x is the number of high-index layers in each stack (both stacks assumed to have equal number of layers); and m is the order of the stack. For a low-refractive index spacer layer, the bandwidth of all-dielectric Fabry-Perot filter will be

$$\frac{\Delta\lambda_L}{\lambda_0} = \frac{4n_L^{2x-1} n_s}{m\pi n_H^{2x}} \qquad (3.28)$$

From (3.27) and (3.28) we can conclude that the bandwidth of the all-dielectric Fabry-Perot filter decreases with the number of layers in the stacks with the index contrast between the low- and high-refractive index layers in the stack. As an illustration, in figure 3.22 the numerically calculated transmittance spectra of titanium dioxide/magnesium fluoride narrowband-pass Fabry-Perot filters on glass substrate are given for 12, 14, and 16 layers in the stacks.

The central position of the transmittance peak in the all-dielectric Fabry-Perot filters is the same as for general Fabry-Perot filters (3.25). However, the effect of variations in angle of incidence is more severe for narrowband-pass filters than that of bandpass, band edge or multilayer reflectors due to generally narrow transmittance peak (see (3.27) and (3.28)). As an example, figure 3.23 shows the numerically calculated TE-polarization transmittance spectra of titanium dioxide/magnesium fluoride narrowband-pass Fabry-Perot filters having a total of 17 layers on glass substrate for different angles of incidence. The shift of the central position of the transmittance peak as strong as 800% in terms of the transmittance peak half width at a 30 deg. Angle of incidence is demonstrated. Such a strong angular dependence of the transmittance spectra of all-dielectric Fabry-Perot filters causes the strong dependence of the transmittance spectra on the convergence (or divergence) of the incident beam. As an example, figure 3.24 shows the numerically calculated TE-polarized transmittance spectra of titanium dioxide/magnesium fluoride narrowband-pass Fabry-Perot filter having a total of 17 layers on glass substrate for different diversions of normal incident Gaussian beam. One can see that the performance of such a filter quickly degrades with the increase on the convergence (or divergence) of the

Gaussian beam. In fact, all-dielectric Fabry-Perot filters are suitable only for plane-parallel or small convergent (or divergent) beams, which cause additional complexity of the optical schemes, employed such a filters.

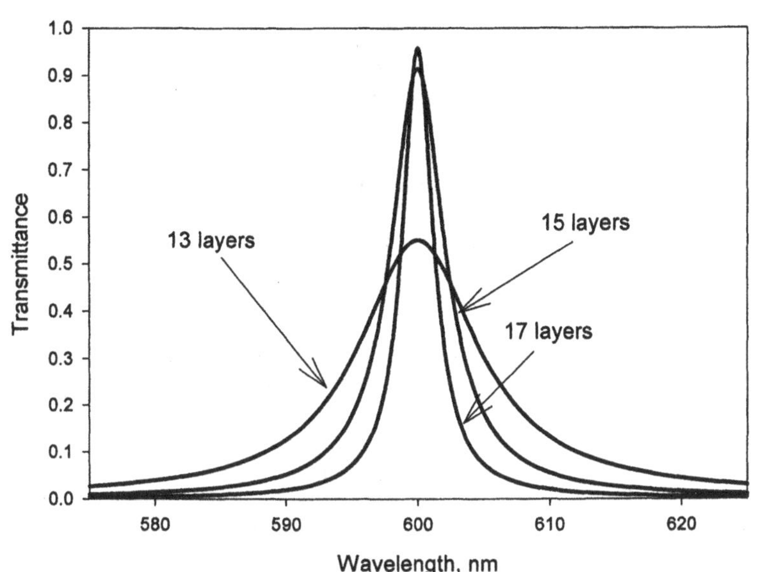

Figure 3.22. Numerically calculated transmittance spectra of titanium dioxide/magnesium fluoride narrowband-pass Fabry-Perot filters having 13, 15, and 17 layers on glass substrate.

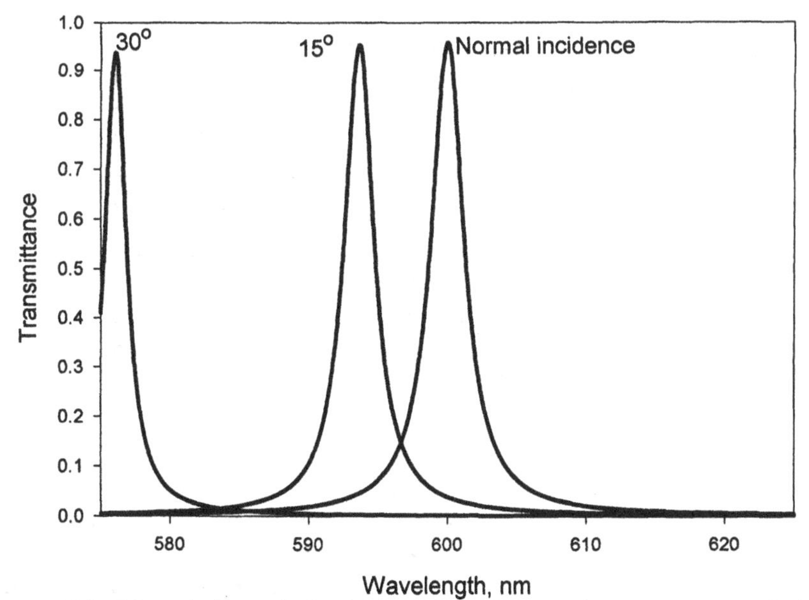

Figure 3.23. Numerically calculated TE-polarized transmittance spectra of titanium dioxide/magnesium fluoride narrowband-pass Fabry-Perot filters having a total of 17 layers on glass substrate for different angles of incidence.

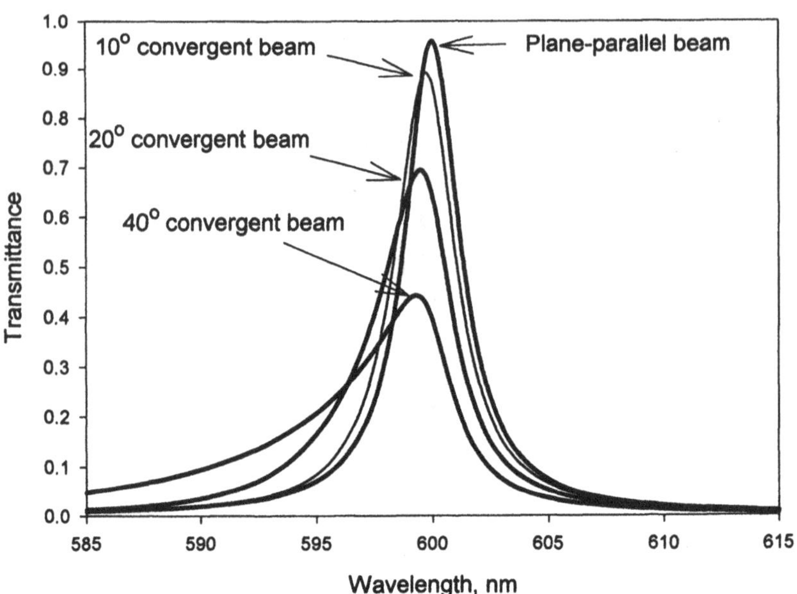

Figure 3.24. Numerically calculated TE-polarized transmittance spectra of titanium dioxide/magnesium fluoride narrowband-pass Fabry-Perot filter having a total of 17 layers on glass substrate for different diversions of incident beam.

3.6.3 Multiple-Cavity Narrowband-Pass Filters.

As was mentioned above and clearly illustrated by figure 3.22, the transmittance spectral shape of the all-dielectric Fabry-Perot filter is not ideal. For filter purposes a nearly rectangular shape of transmittance spectra is desired. In addition, as discussed above, the maximum achievable rejection in the rejection zone of the filter and the bandwidth of transmission zone are related. That is, for a given rejection the value of bandwidth of the filter is predetermined if the refractive indices of the layers in the filter are fixed. The solution of this problem was found in using *multiple-cavities* filter designs. I'm not intending to give here a thorough description of this topic but rather provide some examples with brief explanations. For more detailed information readers can refer to [34].

The simplest type of multiple cavities filters is the double-cavity filter. Such a filter has the structure of *{Air | reflector | half-wave spacer | reflector | half-wave spacer | reflector | Substrate}*. An example of the numerically calculated transmittance spectra of titanium dioxide/magnesium fluoride multilayer on the glass substrate narrowband-pass Fabry-Perot filters having one and two cavities is given in figure 3.25.

The advantages of double-cavity design with respect to single-cavity design are obvious from figure 3.25. However, for some applications, like dense wavelength division multiplexing (DWDM), better spectral shape is needed. The important criteria in high-performance narrowband-pass filters are steeper edges and a flatter top of the transmission peak. For two-cavity filter design the peaks at both sides of pass band (so-called *rabbit's ears*) are prominent. In this case the number of cavities used can be considerably more than two.

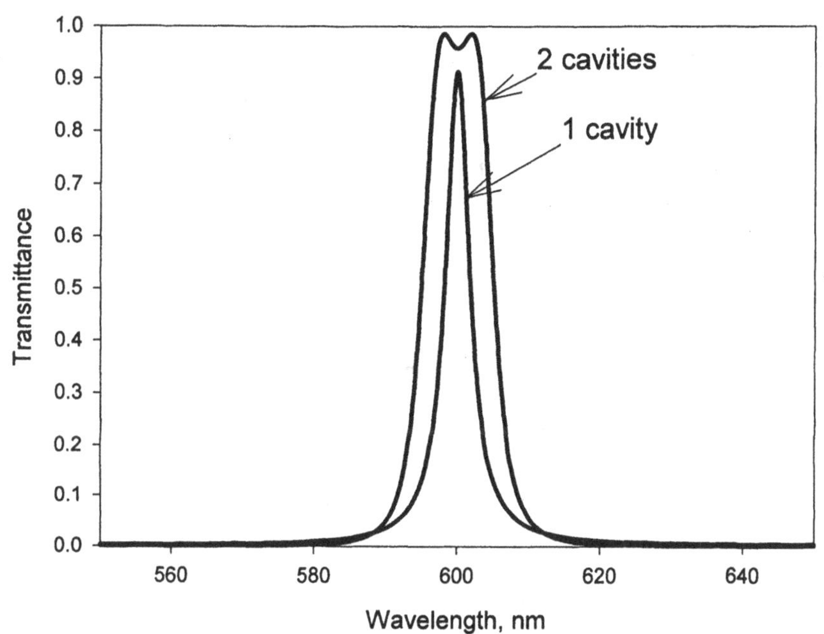

Figure 3.25. Numerically calculated transmittance spectra of titanium dioxide/magnesium fluoride multilayer on the glass substrate narrowband-pass Fabry-Perot filters having one and two cavities (19 layers for single cavity and 25 layers for double cavity {*Air* | *H L H L H L HH L H L H L H L H L H L HH L H L H L H* | *Glass*}).

As an example of an optimized filter structure with strongly suppressed rabbit ears, Figure 3.26 shows the numerically calculated transmittance spectrum of a 70-layer seven-cavity all-dielectric narrowband-pass filter at normal incidence in linear (a) and logarithmic (b) scales.

It should be noted that although at normal incidence the advantages of multiple-cavity filters are obvious, the effects of variations of angle of incidence and beam diversion on the transmittance spectra are more severe for multiple-cavity all-dielectric Fabry-Perot filters than of the single cavities ones. It happens since the rectangular shape of the pass band of the multiple-cavity filter is due to phase matching between the waves, reflected from the different reflector stacks in a multiple-cavity structure that hold only for some distinct angle and wavelength. In figure 3.27a the numerically calculated TE-polarized transmittance spectra for the filter of figure 3.26 is given for different angles of incidence.

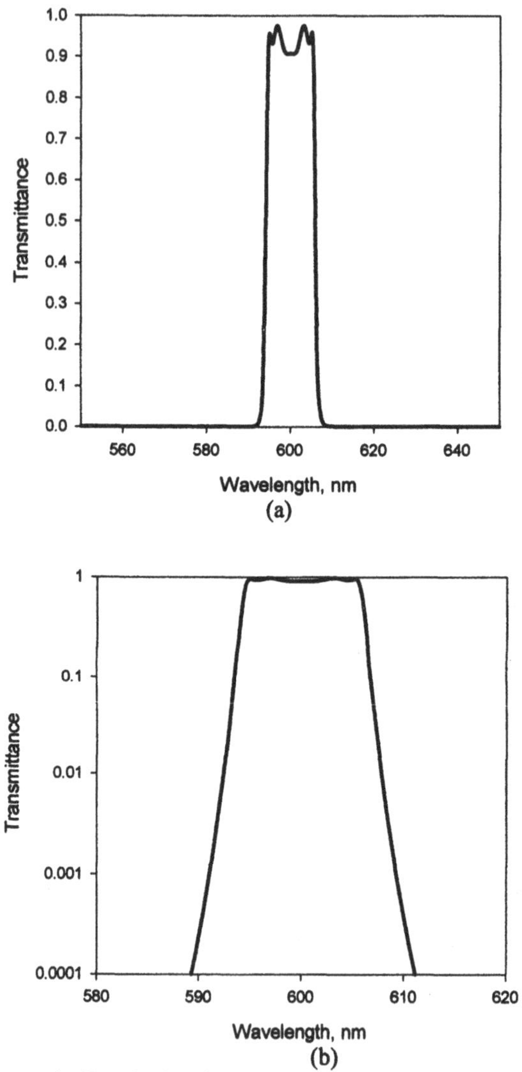

Figure 3.26. The numerically calculated transmittance spectrum of a 70-layer seven-cavity all-dielectric narrowband-pass filter at normal incidence in linear (a) and logarithmic (b) scales.

Unlike the single-cavity filters, where the transmittance peak experiences wavelength shift when illuminated at the not-normal angles without significant perturbation of its shape, in multiple-cavity all-dielectric Fabry–Perot filters the shape of the transmittance band changes dramatically with variations of angle of incidence. In fact, the flat top of the multiple-cavity filter at normal incidence transforms into separate narrow transmittance peaks related to the interference between the waves reflected from different reflector stacks within the multiple-cavity multilayer structure. Hence, multiple cavity all-dielectric Fabry-Perot filters became unusable at the angles spaced more than 3 to 5 degrees apart from normal incidence direction.

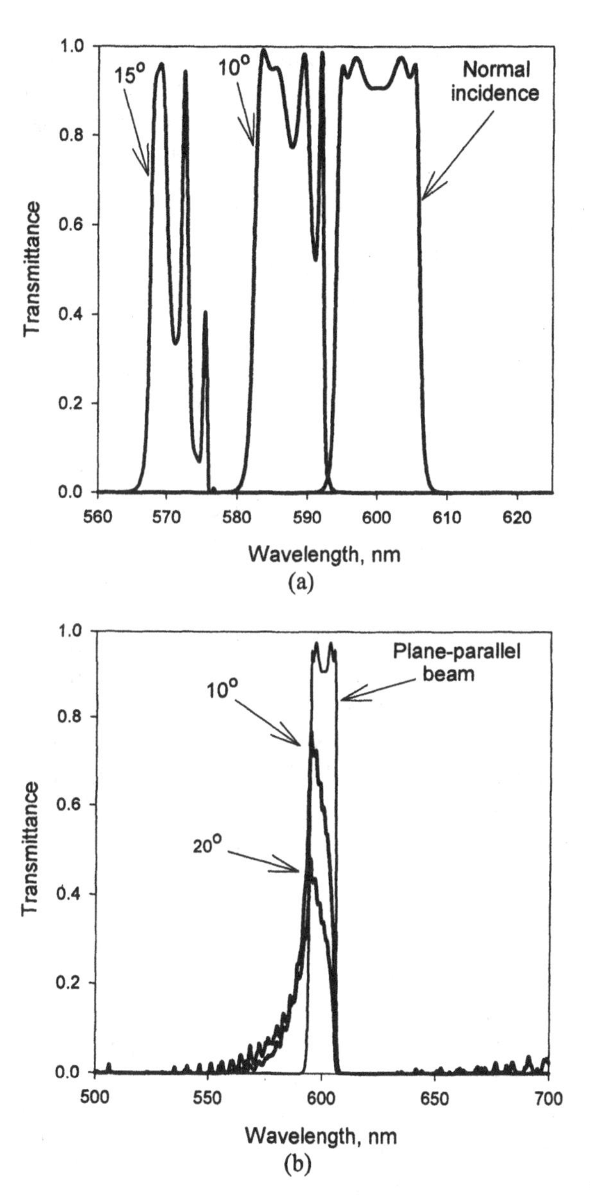

Figure 3.27. Numerically calculated TE-polarized transmittance spectra for the filter of figure 3.26 for different angles of incidence (a) and for different diversions of incident at normal incidence Gaussian beam (b).

Such a property is extremely important in DWDM filters and requires several hundred layers to produce flat-top transmittance bands with bandwidth below 1 nm. Additional precise mechanical alignment usually solves this problem.

In addition, multiple-cavity all-dielectric Fabry-Perot filters require high collimation of the incident beam. In figure 3.27b the numerically calculated TE-polarized transmittance spectra for the filter of figure 3.26 is given for different diversions of incident at a normal-incidence Gaussian beam. One can see that the shape of the pass band of such a filter degrades significantly even for the Gaussian beams, whose convergence (or divergence) angle is about 10 degrees. Hence, multiple-cavity all-dielectric Fabry-Perot filters require not only precise mechanical alignment to ensure normal incidence of the beam but also a high degree of collimation.

3.6.4 Multiple-Cavity Metal-Dielectric Filters

Multiple-cavity all-dielectric Fabry-Perot filters, discussed in section 3.6.3, have several significant disadvantages. These disadvantages include the presence of the long-wave pass bands (i.e., wavelength-limited rejection bands) and significant difficulties in manufacturing such filters for short spectral range (deep and far ultraviolet). For the applications that require the above filter properties, *multiple-cavity metal dielectric filters* are usually used. Single-cavity metal-dielectric Fabry-Perot filters are discussed in section 3.6.1 of this book. In particular, it was found that in addition to the disadvantages of an all-dielectric Fabry-Perot filter like the relation between pass-band bandwidth and maximum obtainable rejection and triangular shape of the pass band, the single-cavity metal-dielectric Fabry-Perot filters exhibit enhancement of losses with the decrease of the pass-band bandwidth due to the losses in metal.

In multiple-cavity metal-dielectric Fabry-Perot filters this problem is usually solved by the induced-transmission design. The phenomenon serves as the base of such filters so that it is possible to match metal layers and dielectric spacers thickness such that for given wavelength and angle of incidence the localization of the light in the metal layers during transmission is minimal while maximal inside the dielectric layers. Using such design, it is still impossible to achieve perfect transmission through such a filter. However, the transmission can be made above 50%, combined with near square pass band shape and good control of rejection and pass band bandwidth simultaneously.

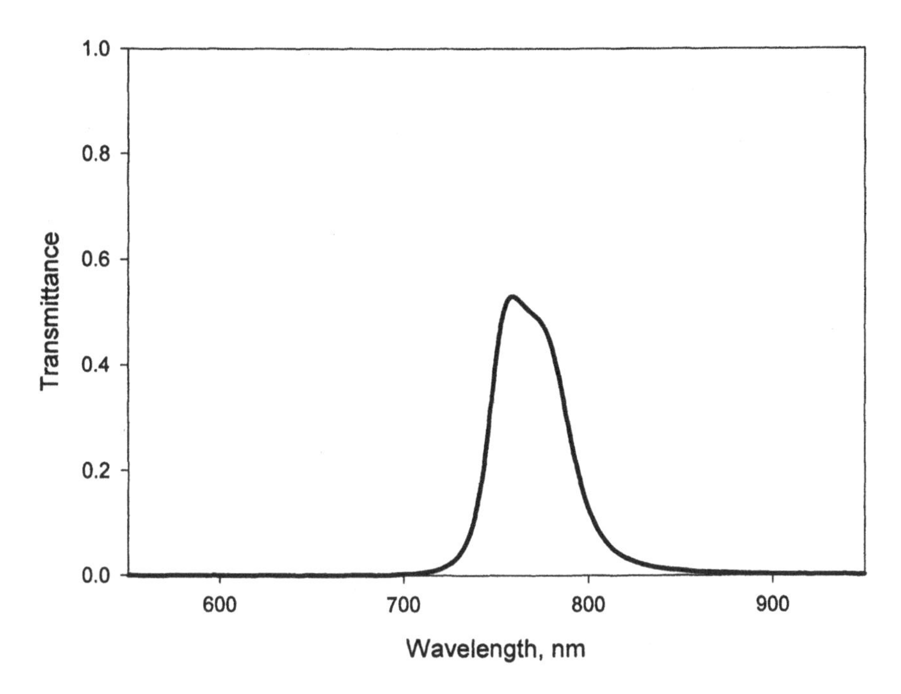

Figure 3.28. A numerically calculated transmittance spectrum of a metal dielectric multiple-cavity filter at normal incidence. The structure of the filter was {Glass | 40 nm of silver | 210 nm of magnesium fluoride | 40 nm of silver | 150 nm of magnesium fluoride | 40 nm of silver | 210 nm of magnesium fluoride | 40 nm of silver | Glass}.

As an example, in figure 3.28 the numerically calculated transmittance spectrum of metal dielectric multiple-cavity filter at normal incidence is presented for the following filter structure: {Glass | 40 nm of silver | 210 nm of magnesium fluoride | 40 nm of silver | 150 nm of magnesium fluoride | 40 nm of silver | 210 nm of magnesium fluoride | 40 nm of silver | Glass}. Note that the total thickness of silver in such a filter is 120 nm. The transmission through 120 nm silver film will be ~ 0.001%.

Multiple-cavity metal-dielectric filters have some significant disadvantages. In addition to an angular shift of the wavelength position of the pass band, due to absorption in metal layer, such filters are not suitable for high-power applications. The temperature dependence of the optical performance of such filters is also the strongest among all interference-based filters.

Chapter 4

OMNIDIRECTIONAL DIELECTRIC REFLECTORS

4.1 Introduction

Classic multilayer dielectric reflectors are considered in section 3.3.2 of this book. In is shown that the classic designs of such mirrors have some disadvantages despite very low losses, like a limited wavelength range of reflection bands and the dependence of the spectral position of the reflection band on the angle of incidence. However, very recently it was found ([8]), that by using different design rules, it is possible to create a multilayer dielectric structure that offers metallic-like omnidirectional reflectivity together with frequency selectivity and low-loss behavior typical for multilayer dielectric reflectors. This chapter is devoted to the discussion of such a multilayer structure. To understand how such a structure works, it will be useful to understand Bloch waves and band structures of periodic multilayer media.

4.2 Bloch Waves and Band Structures

Reflection from and transmission through the multilayer stack are discussed in sections 2.5, 2.9, and 3.4 of this book. However, the multilayer stack is considered in those sections as a black box having outputs related to inputs without developing the understanding of how the light propagates though such a structure. Several methods have been developed to describe behavior of a lightwave propagating through periodic media. The approach that is presented here is the *Bloch wave* approach. I follow the description given by P. Yeh [21].

Let us consider the infinite periodic multilayer consisting of isotropic homogeneous layers having plane interfaces with an index of refraction profile (see figure 4.1)

$$n(x) = \begin{cases} n_2, & 0 < x < b; \\ n_1, & b < x < \Lambda. \end{cases} \tag{4.1}$$

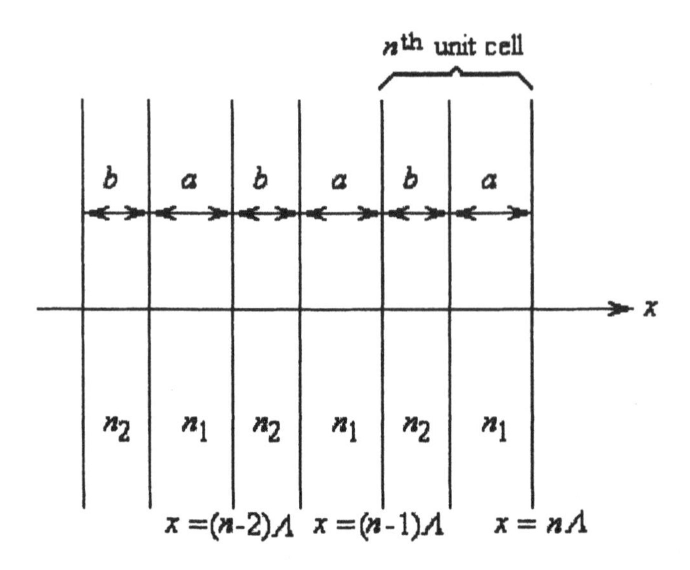

Figure 4.1. A schematic drawing of a periodic multilayer medium.

$\Lambda = a + b$ is called the period of such a periodic multilayer medium, since

$$n(x) = n(x + \Lambda) \tag{4.2}$$

The periodic multilayer medium is assumed to be infinite. Hence, according to the Floquet theorem, solutions of the wave equations (2.8) and (2.9) have the form

$$E_K(x, z) = E_K(x)e^{-i\beta z}\, e^{-iKx} \tag{4.3}$$

where $E_K(x)$ is periodic function with the period Λ:

$$E_K(x) = E_K(x + \Lambda) \tag{4.4}$$

From another point of view, the electric field in the nth unit cell can be written in the following form (see section 2.5):

$$E(x) = \begin{cases} a_n e^{-i\cdot k_{1x}(x-n\Lambda)} + b_n e^{i\cdot k_{1x}(x-n\Lambda)} & n\Lambda - a < x < n\Lambda \\ c_n e^{-i\cdot k_{2x}(x-n\Lambda+a)} + d_n e^{i\cdot k_{1x}(x-n\Lambda+a)} & (n-1)\Lambda < x < n\Lambda - a \end{cases} \tag{4.5}$$

where $k_{1x} = \sqrt{\left(\frac{n_1 \omega}{c}\right)^2 - \beta^2}$, $k_{1x} = \sqrt{\left(\frac{n_1 \omega}{c}\right)^2 - \beta^2}$. From the boundary conditions (2.48) to (2.51) it can be derived that the amplitude coefficients a_n, b_n, c_n, and d_n for TE-polarized waves are connected through the following relations (the exact derivation can be found in [21]):

$$\begin{pmatrix} a_{n-1} \\ b_{n-1} \end{pmatrix} = \frac{1}{2} \begin{pmatrix} e^{i \cdot k_{2x} b}\left(1 + \frac{k_{2x}}{k_{1x}}\right) & e^{-i \cdot k_{2x} b}\left(1 - \frac{k_{2x}}{k_{1x}}\right) \\ e^{i \cdot k_{2x} b}\left(1 - \frac{k_{2x}}{k_{1x}}\right) & e^{-i \cdot k_{2x} b}\left(1 + \frac{k_{2x}}{k_{1x}}\right) \end{pmatrix} \begin{pmatrix} c_n \\ d_n \end{pmatrix} \quad (4.6)$$

$$\begin{pmatrix} c_n \\ d_n \end{pmatrix} = \frac{1}{2} \begin{pmatrix} e^{i \cdot k_{1x} a}\left(1 + \frac{k_{1x}}{k_{2x}}\right) & e^{-i \cdot k_{1x} a}\left(1 - \frac{k_{1x}}{k_{2x}}\right) \\ e^{i \cdot k_{1x} a}\left(1 - \frac{k_{1x}}{k_{2x}}\right) & e^{-i \cdot k_{1x} a}\left(1 + \frac{k_{1x}}{k_{2x}}\right) \end{pmatrix} \begin{pmatrix} a_n \\ b_n \end{pmatrix} \quad (4.7)$$

From (4.6) and (4.7), one can obtain following matrix equation:

$$\begin{pmatrix} a_{n-1} \\ b_{n-1} \end{pmatrix} = \frac{1}{2} \begin{pmatrix} A & B \\ C & D \end{pmatrix} \begin{pmatrix} a_n \\ b_n \end{pmatrix} \quad (4.8)$$

where

$$A = e^{i \cdot k_{1x} a}\left[\cos k_{2x} b + \frac{1}{2} i \left(\frac{k_{2x}}{k_{1x}} + \frac{k_{1x}}{k_{2x}}\right) \sin k_{2x} b\right]$$

$$B = e^{-i \cdot k_{1x} a}\left[\frac{1}{2} i \left(\frac{k_{2x}}{k_{1x}} + \frac{k_{1x}}{k_{2x}}\right) \sin k_{2x} b\right]$$

$$C = e^{i \cdot k_{1x} a}\left[-\frac{1}{2} i \left(\frac{k_{2x}}{k_{1x}} - \frac{k_{1x}}{k_{2x}}\right) \sin k_{2x} b\right]$$

$$D = e^{-i \cdot k_{1x} a}\left[\cos k_{2x} b - \frac{1}{2} i \left(\frac{k_{2x}}{k_{1x}} + \frac{k_{1x}}{k_{2x}}\right) \sin k_{2x} b\right] \quad (4.9)$$

For TM waves the coefficients in the matrix equation (4.8) are

$$A_{TM}=e^{i\cdot k_{1x}a}\left[\cos k_{2x}b+\frac{1}{2}i\left(\frac{n_2^2\cdot k_{1x}}{n_1^2\cdot k_{2x}}+\frac{n_1^2\cdot k_{2x}}{n_2^2\cdot k_{1x}}\right)\sin k_{2x}b\right]$$

$$B_{TM}=e^{-i\cdot k_{1x}a}\left[\frac{1}{2}i\left(\frac{n_2^2\cdot k_{1x}}{n_1^2\cdot k_{2x}}-\frac{n_1^2\cdot k_{2x}}{n_2^2\cdot k_{1x}}\right)\sin k_{2x}b\right]$$

$$C_{TM}=e^{i\cdot k_{1x}a}\left[-\frac{1}{2}i\left(\frac{n_2^2\cdot k_{1x}}{n_1^2\cdot k_{2x}}-\frac{n_1^2\cdot k_{2x}}{n_2^2\cdot k_{1x}}\right)\sin k_{2x}b\right]$$

$$D_{TM}=e^{-i\cdot k_{1x}a}\left[\cos k_{2x}b-\frac{1}{2}i\left(\frac{n_2^2\cdot k_{1x}}{n_1^2\cdot k_{2x}}+\frac{n_1^2\cdot k_{2x}}{n_2^2\cdot k_{1x}}\right)\sin k_{2x}b\right] \tag{4.10}$$

On the other hand, as it follows from (4.4), for a periodic multilayer

$$\begin{pmatrix}a_n\\ b_n\end{pmatrix}=e^{-iK\Lambda}\begin{pmatrix}a_{n-1}\\ b_{n-1}\end{pmatrix} \tag{4.11}$$

That is, according to (4.8) and (4.11):

$$\begin{pmatrix}A & B\\ C & D\end{pmatrix}\begin{pmatrix}a_n\\ b_n\end{pmatrix}=e^{iK\Lambda}\begin{pmatrix}a_n\\ b_n\end{pmatrix} \tag{4.12}$$

Equation (4.12) is an eigenvalue problem, and $e^{iK\Lambda}$ is the eigenvalue of matrix (A, B, C, D):

$$e^{iK\Lambda}=\frac{1}{2}(A+D)\pm\sqrt{\left[\frac{1}{2}(A+D)\right]^2-1} \tag{4.13}$$

Hence, the Bloch waves are the eigenvectors of matrix (A, B, C, D) with eigenvalues $e^{\pm iK\Lambda}$ (the sign corresponds to forward or backward propagating Bloch waves), given by (4.13). As follows from (4.13), the dispersion relation between ω, β and K has the form

$$K(\beta,\omega)=\frac{1}{\Lambda}\operatorname{acos}\left[\frac{1}{2}(A+D)\right] \tag{4.14}$$

By substituting (4.9) and (4.10) into (4.14), one can obtain

$$K_{TE}(\beta,\omega)=\frac{1}{\Lambda}\operatorname{acos}[\cos k_{1x}a\cdot\cos k_{2x}b-$$

$$-\frac{1}{2}\left(\frac{k_{1x}}{k_{2x}}+\frac{k_{2x}}{k_{1x}}\right)\sin k_{1x}a\cdot\sin k_{2x}b] \quad (4.15a)$$

$$K_{TM}(\beta,\omega)=\frac{1}{\Lambda}\operatorname{acos}[\cos k_{1x}a\cdot\cos k_{2x}b-$$

$$-\frac{1}{2}\left(\frac{n_2^2\cdot k_{1x}}{n_1^2\cdot k_{2x}}+\frac{n_1^2\cdot k_{2x}}{n_2^2\cdot k_{1x}}\right)\sin k_{1x}a\cdot\sin k_{2x}b] \quad (4.15b)$$

One can see from (4.15a-b) that Bloch wave propagation constant K can take both real and imaginary values depending on the particular value of expressions in "[]" branches. Real values of K correspond to propagating Bloch waves, while imaginary values of K correspond to evanescent (i.e. damped) Bloch waves. The regions on the (β, ω) plane where Bloch waves are propagating (i.e., Bloch wave propagation constant is purely real) are called *pass-bands*, while the regions on the (β, ω) plane where Bloch waves are evanescent (i.e., Bloch wave propagation constant has nonzero imaginary part) are called *stop-bands* or *forbidden bands*. From (4.15) it is obvious that the band edges are the solutions of following equations:

TE waves: $$[\cos k_{1x}a\cdot\cos k_{2x}b-$$

$$-\frac{1}{2}\left(\frac{k_{1x}}{k_{2x}}+\frac{k_{2x}}{k_{1x}}\right)\sin k_{1x}a\cdot\sin k_{2x}b]=1 \quad (4.16a)$$

TM waves: $$[\cos k_{1x}a\cdot\cos k_{2x}b-$$

$$-\frac{1}{2}\left(\frac{n_2^2\cdot k_{1x}}{n_1^2\cdot k_{2x}}+\frac{n_1^2\cdot k_{2x}}{n_2^2\cdot k_{1x}}\right)\sin k_{1x}a\cdot\sin k_{2x}b]=1$$

(4.16b)

As an example, figure 4.2 gives the numerically calculated band for a silicon dioxide/titanium dioxide quarter wave stack for TE (right half) and TM (left half) waves. The shaded zones are allowed bands; the parallel wavevector has the dimensions of $4\pi/\Lambda$, and the frequency has the dimensions of $4\pi c/\Lambda$. The TM forbidden band degenerates exactly at the Brewster angle conditions (see equation (2.35) from section 2.3) due to the fact that, at this angle, the incident and reflected waves are uncoupled. Waves originating from the medium outside the multilayer satisfy the condition $\omega \geq c\beta/n_0$, where n_0 is the refractive index of the outside medium. Therefore, the region within the (β, ω) plane, where Bloch waves can be excited at the edge of such a multilayer, is above the *light line* ($\omega = c\beta/n_0$). Bloch waves with a parallel wavevector equal to zero are excited by normally incident on multilayer light. Bloch waves, whose propagation constants are lying on the light line, are excited by the light incident on the multilayer at the angle of 90°. In the regions of the (β, ω) plane outside the light lines cone, the Bloch waves can be excited only indirectly (for example, through broken internal reflection).

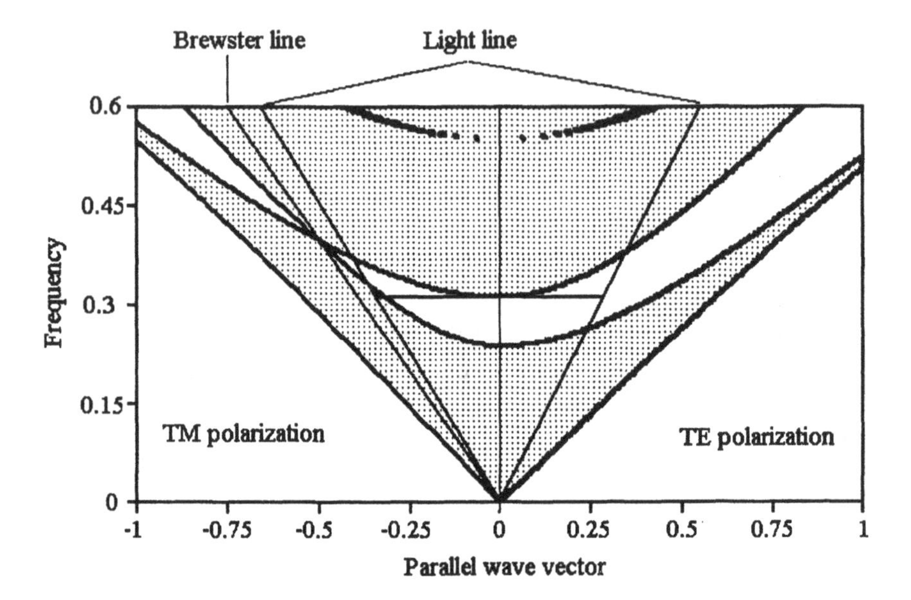

Figure 4.2. Numerically calculated band structure of dielectric multilayer consisting of alternating layers of silicon dioxide and titanium dioxide (n = 1.5 and 2.3, respectively), with thicknesses of 166.7 and 113.6 nm, respectively. The shaded zones are allowed bands; parallel wavevector has the dimensions of $4\pi/\Lambda$ and frequency has the dimensions of $4\pi c/\Lambda$.

Assuming that the layers comprising the multilayer medium are lossless, the multilayer is transparent within the pass-bands. Within the forbidden band, the multilayer medium becomes reflective. In the effective medium approximation, one can estimate the decrease of transmission through the multilayer medium of thickness l as $e^{-l\mathrm{Im}[K]}$ (this approximation is valid when l considerably exceeds Λ). Hence, as the thickness of the multilayer (i.e., the number of layers) is increased, the transmission decreases exponentially, while the reflectivity approaches unity. The values of Im[K] are not uniform across the forbidden bands. They are maximal in the centers of the forbidden zones and minimal around the band edges. As an example, figure 4.3 gives the numerically calculated contour plot of *log*[Im(K)] for the dielectric multilayer of figure 4.2.

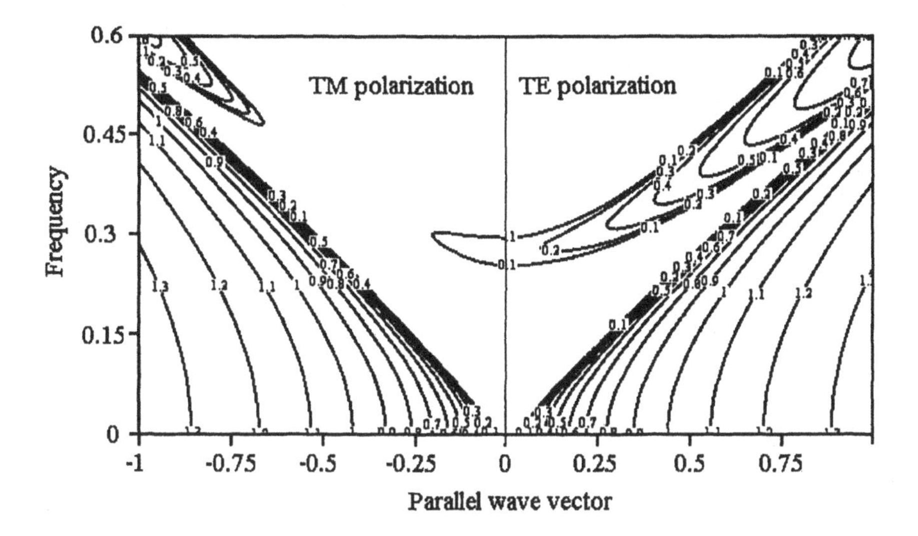

Figure 4.3. The numerically calculated contour plot of *log*[Im(*K*)] for the dielectric multilayer offFigure 4.2.

4.3 A Dielectric Omnidirectional Reflector

As one can see from figure 4.2, there is no frequency (i.e., wavelength) for which a titanium dioxide/silicon dioxide quarter-wave stack is reflective for any incident angle (in other words, within the light cone). For anyfrequency one can always find such a value of a parallel wavevector for which a Bloch wave at said frequency can propagate in the crystal and, hence, transmit through the multilayer. As was pointed out in [8], the necessary and sufficient criterion for omnidirectional reflectivity of a dielectric multilayer at a given frequency is that no transmitting state of the structure exists inside the light cone. More particularly, a necessary condition for omnidirectional reflectivity is that the light line is above the Brewster line of the multilayer medium because at Brewster angle, the TM mode will be transmitted through the multilayer (noted above degeneration of the forbidden bands). This condition is satisfied if refractive indices of the layers comprising the multilayer (n_1 and n_2) and that of outside medium satisfy following relation:

$$\mathrm{asin}(n_0/\min\{n_1,n_2\}) < \theta_B.$$

A sufficient condition for omnidirectional reflectivity is the existence of a particular frequency at which no propagating mode within the crystal exists above the light cone (see figure 4.2). For example, the silicon dioxide/titanium dioxide quarter-wave stack of figure 4.2 does not have an omnidirectional reflectivity range even though its Brewster crossing lies outside the light cone. In [8] it was pointed out that this is due to the large group velocity of modes in the lower band edge of the TM mode that allows every frequency to couple to a propagating state in the crystal.

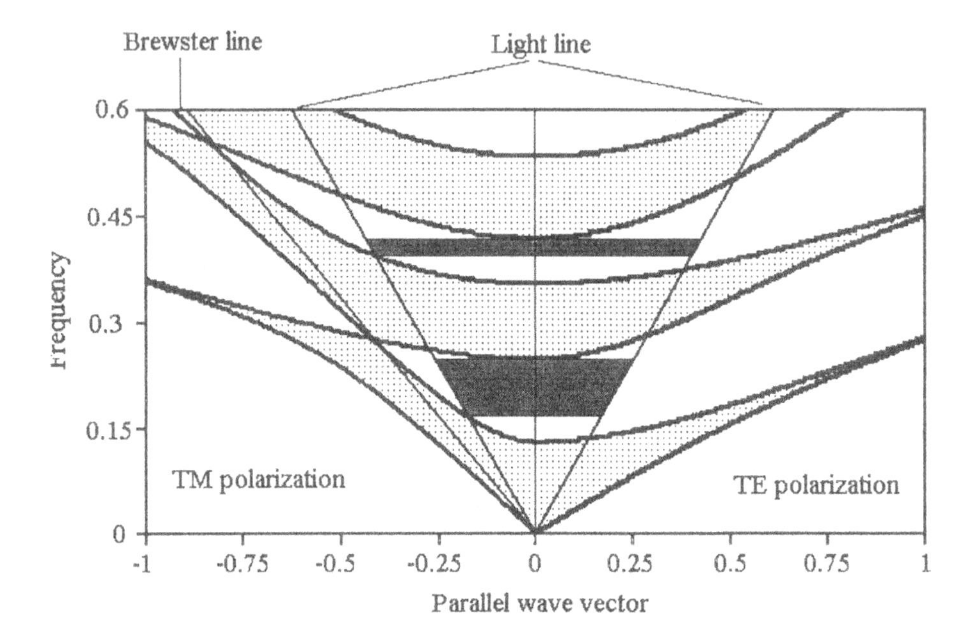

Figure 4.4. The numerically calculated band structure of the dielectric multilayer consisting of alternating layers of polystyrene and tellurium (n = 1.6 and 4.6, respectively), with thickness ratio of 1.6/0.8 (as was used in [8]'s experiment). The shaded zones are allowed bands; the parallel wavevector has the dimensions of $4\pi/\Lambda$, and frequency has the dimensions of $4\pi c/\Lambda$.

One can expect that, in order to get omnidirectional reflectivity, the refractive index contrast ($|n_1 - n_2|/\min\{n_1,n_2\}$) of layers comprising the multilayer should be enlarged, while keeping a low refractive index layer in the multilayer ($\min\{n_1,n_2\}$) at sufficiently low range to meet the Brewster condition. Indeed, in [8] it was demonstrated that the multilayer, consisting of alternating layers of polystyrene (n = 1.6) and tellurium (n = 4.6), exhibits an omnidirectional reflection band. In figure 4.4 the numerically calculated band structure of the dielectric multilayer, consisting of alternating layers of polystyrene and tellurium, with thickness ratio of 1.6/0.8 (as was used in the [8] experiment) is given.

One can see from figure 4.4 that for a multilayer structure, exploited by [8], two frequency ranges of omnidirectional reflection exist, corresponding to the first and second forbidden bands of the multilayer. Both omnidirectional reflection bands are limited from above by the normal incidence band edge

ω_h and from below by the intersection of the top of the TM allowed band edge with the light line ω_l . It is convenient to quantify the width of the omnidirectional frequency range by dimensionless parameter *range to midrange ratio*, which can be defined as $(\omega_h - \omega_l)/1/2(\omega_h + \omega_l)$.

From (4.16) we can estimate the width of the forbidden band for a given incident angle θ_0:

$$\Delta\omega(\theta_0) = \frac{2c}{a\sqrt{n_1^2 - n_0^2 \sin^2\theta_0} + b\sqrt{n_2^2 - n_0^2 \sin^2\theta_0}} \times$$
$$\times \left[\operatorname{acos}\left(-\sqrt{\frac{\Gamma - 1}{\Gamma + 1}} \right) - \operatorname{acos}\left(\sqrt{\frac{\Gamma - 1}{\Gamma + 1}} \right) \right] \tag{4.17}$$

where

$$\Gamma \equiv \begin{cases} \frac{1}{2}\left(\frac{k_{2x}}{k_{1x}} + \frac{k_{1x}}{k_{2x}} \right) & TE \\ \frac{1}{2}\left(\frac{n_1^2 k_{2x}}{n_2^2 k_{1x}} + \frac{n_2^2 k_{1x}}{n_1^2 k_{2x}} \right) & TM \end{cases} \tag{4.18}$$

As follows from (4.17) and as it should be due to symmetry reasons, at normal incidence conditions (parallel wavevector equal to zero) TM and TE mode forbidden bands coincide. As the angle of incidence increases, the forbidden band of TE-polarized Bloch waves increases, while the forbidden band of the TM-polarized Bloch waves decreases (see figures 4.3 and 4.4). For both polarizations, the center of the forbidden band shifts to higher frequencies with the increase in the angle of incidence. Hence, according to [8], the criterion for the existence of omnidirectional reflectivity can be restated as the occurrence of a frequency overlap between the forbidden band at normal incidence and the forbidden band of the TM Bloch wave at 90 deg. Edges of omnidirectional reflection band can be written in the following form:

$$\omega_h = \frac{2c}{n_2 b + n_1 a} \operatorname{acos}\left(-\left| \frac{n_1 - n_2}{n_1 + n_2} \right| \right) \tag{4.19a}$$

$$\omega_l = \frac{2c}{b\sqrt{n_2^2 - n_0^2} + a\sqrt{n_1^2 - n_0^2}} \times$$

$$\times \operatorname{acos}\left(-\left|\frac{n_1^2\sqrt{n_2^2-n_0^2}-n_2^2\sqrt{n_1^2-n_0^2}}{n_1^2\sqrt{n_2^2-n_0^2}+n_2^2\sqrt{n_1^2-n_0^2}}\right|\right) \quad (4.19b)$$

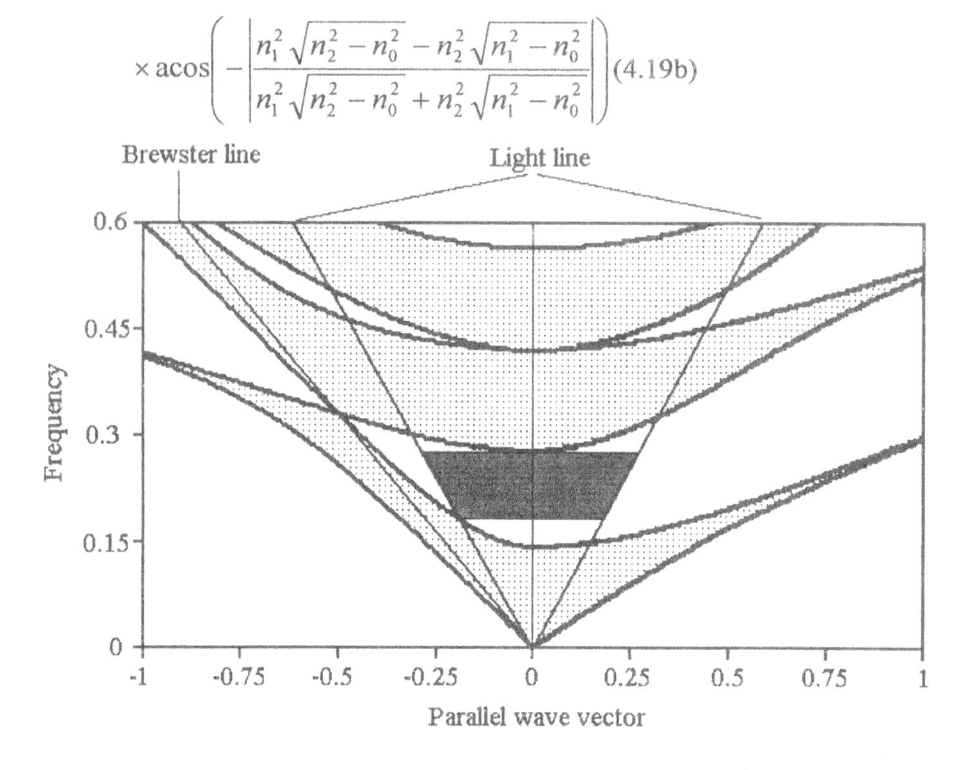

Figure 4.5. Numerically calculated band structure of dielectric multilayer consisting of alternating layers of polystyrene and tellurium (n = 1.6 and 4.6, respectively), with thickness ratio of 2.875 (quarter wave stack). The shaded zones are allowed bands; the parallel wavevector has the dimensions of 4π/Λ, and frequency has the dimensions of 4πc/Λ.

Surprisingly, as follows from (4.19a b) and as was shown in [43], the maximum range to midrange ratio takes place for layers with thicknesses that are not equal to the quarter wave stack. However, the increase in omnidirectional reflection band width gained by deviating from the quarter wave stack is in the order of a few percent [43]. The numerically calculated band structure of polystyrene and tellurium quarter wave stack is presented in figure 4.5.

In figure 4.6 the numerically calculated contour plot of range to midrange ratio is given as a function of n_2/n_1 and n_1/n_0, where ω_h and ω_l are determined by solving (4.19), with quarter wave layers thickness. The contours represent equiomnidirectional ranges for different material index parameters.

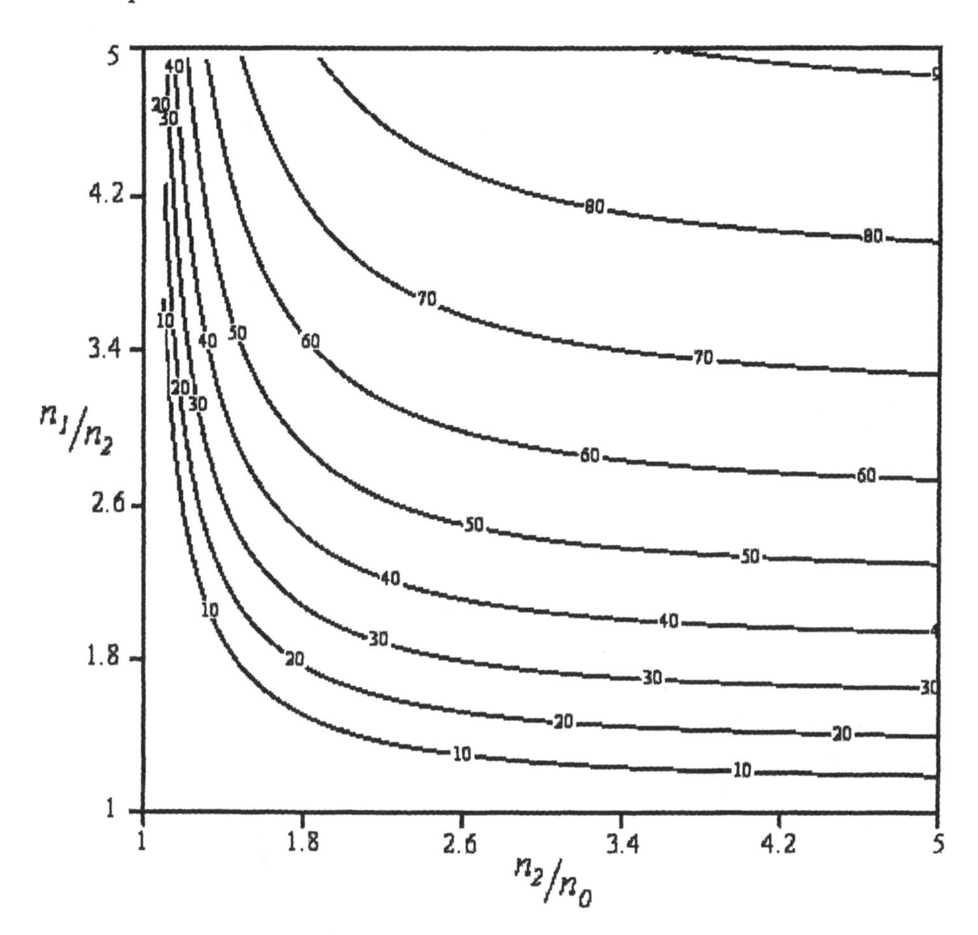

Figure 4.6. The numerically calculated contour plot of range to midrange ratio $(\omega_h - \omega_l)/1/2(\omega_h + \omega_l)$, for the fundamental frequency range of omnidirectional reflection for a quarter- wave stack. It was assumed that $n_l > n_2$.

Figure 4.7 gives the numerically calculated reflectance spectra for the dielectric multilayer of figure 4.4 at normal, 40 deg. and 80 deg. angles of incidence. The multilayer was assumed to have the following structure: {Air | Te | | PS | Te | PS |Te | PS | Te | PS | Te | Glass}(the same as that used by [8]). One can see that the high reflectivity spectral regions at the different angles of incidence overlap, thus forming a reflective spectral region of anyangle of incidence. As for ordinary dielectric reflectors (see section 3.3 of this book), the wavelength location of the omnidirectional range is determined by the layer thickness and can be tuned within the transparency bands of the materials comprising a multilayer. The omnidirectional range for such a structure according to (4.19) is about 5.6 μm, centered at 12.4 μm (i.e., the range to the midrange ratio is about 45%).

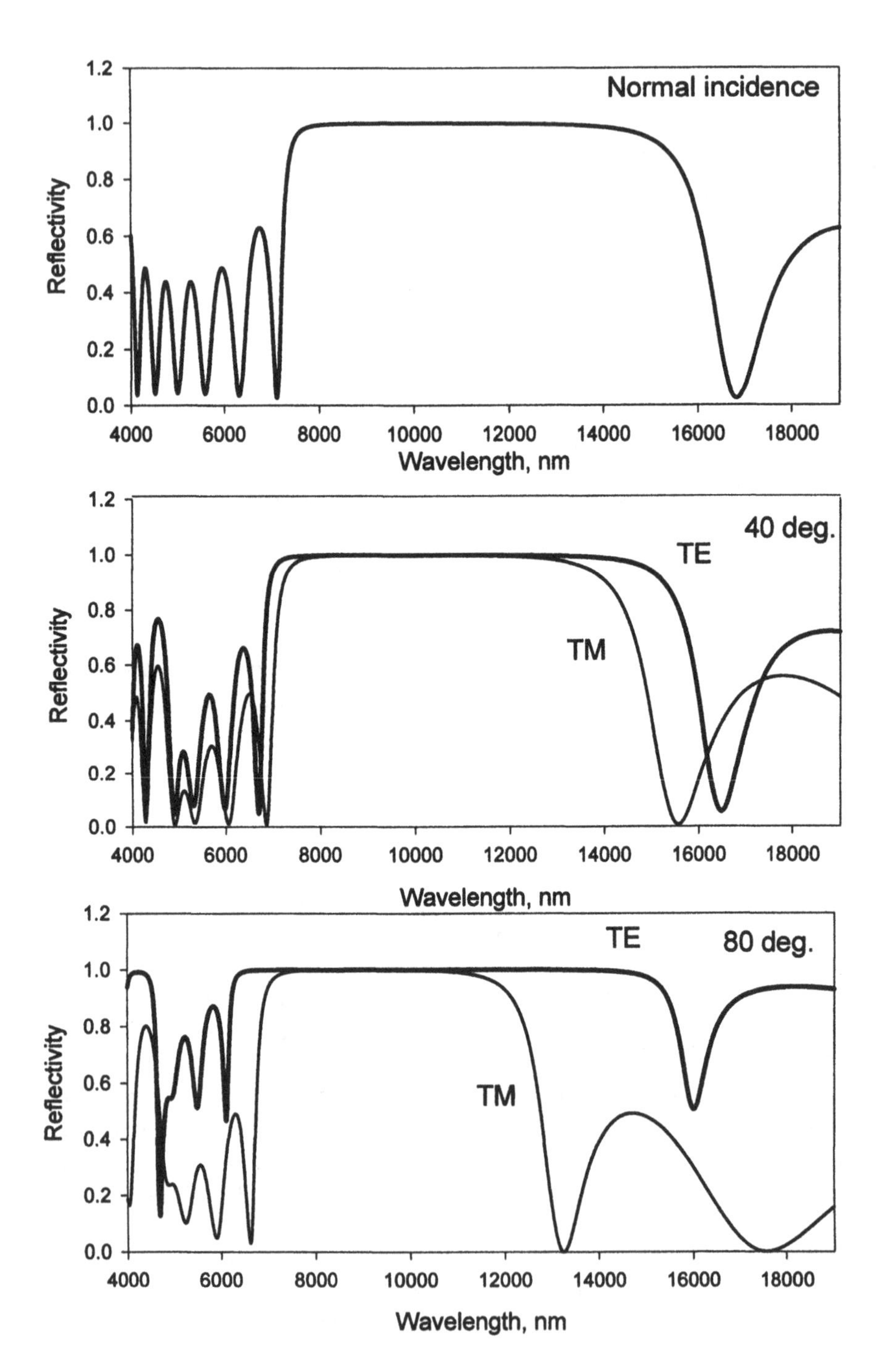

Figure 4.7. Numerically calculated reflectance spectra for the nine-layer dielectric multilayer of figure 4.4 at normal, 40 deg. and 80 deg. angles of incidence.

The pioneering works of [8] and [43] boosted interest in omnidirectional filters, which led to further development of this topic. For example [44] demonstrated the fabrication of an omnidirectional reflector for visible wavelengths. The multilayer of [44] consisted of 19 alternating layers of tin sulfide (n = 2.6) and silica (n = 1.46). Such a reflector showed an omnidirectional range greater than 10% for the visible spectral range. The design for omnidirectional reflectors in the optical telecommunication band is described In [45]. It was found that the high refractive index should be larger than 2.26 and that the low refractive index for maximum omnidirectional bandwidth should be around 1.5. In [46] it was shown that the multilayer should be finished with a low-index layer having a thickness larger than a quarter-wave to increase reflectivity, while layers below may remain of quarter-wave optical thickness at normal incidence angle. In particular, amorphous silicon/silicon dioxide omnidirectional reflectors in the near-infrared range on silicon and silica substrates were demonstrated.

A quite interesting approach for enlargement of omnidirectional reflectivity band for a dielectric multilayer was proposed in [47]. It was theoretically shown that the omnidirectional reflection frequency band can be substantially enlarged by placing together two multilayers having different filling factors. Let us discuss this approach in more detail.

4.4 Enlargement of the Omnidirectional Frequency Band

As mentioned in previous section, it is possible to control the position of the omnidirectional reflection band by modification in the multilayer structure (i.e., layer thickness and refractive indices). It is also possible to control it through modification of the thickness ratio of the layers in the multilayer while keeping the period of the multilayer the same. As an example, let us consider two polystyrene/tellurium multilayers with different thickness ratios. We will follow here the example given in [47].

Figure 4.8 presents the numerically calculated band structure of a dielectric multilayer consisting of alternating layers of polystyrene and tellurium with thickness ratio of 0.8/0.2 (let us denote this structure as DM1). Let us also assume that the medium outside the multilayer is air (n = 1). Figure 4.8 shows the omnidirectional reflection band for DM1 will be in the range of 0.196 to 0.296 $\omega a/2\pi c$, with the range to midrange ratio about 40%. If we set the thickness ratio as 0.55/0.45 for the polystyrene/tellurium multilayer (which is denoted as DM2), its band structure will be like that shown in figure 4.9.

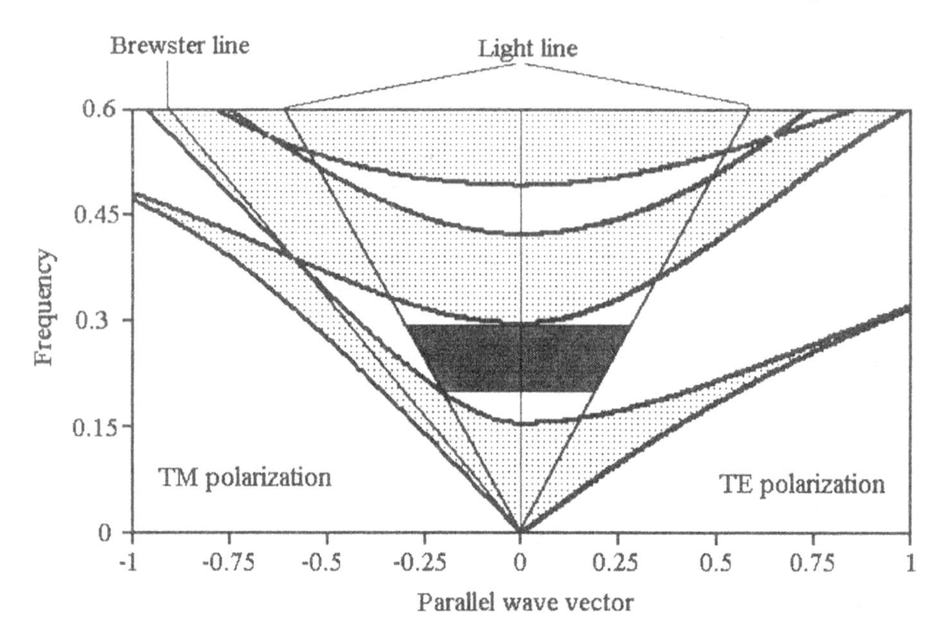

Figure 4.8. The numerically calculated band structure of a dielectric multilayer consisting of alternating layers of polystyrene and tellurium, with a thickness ratio of 0.8/0.2. The shaded zones are the allowed bands, the parallel wavevector has the dimensions of $4\pi/\Lambda$, and frequency has the dimensions of $4\pi c/\Lambda$.

As one can see from figure 4.9, for DM2 we have two omnidirectional reflection bands, 0.146 to 0.208 $\omega a/2\pi c$ and 0.323 to 0.393 $\omega a/2\pi c$, respectively, with the range to midrange ratios about 35% and 20%, respectively. According to figures 4.8 and 4.9, both omnidirectional reflection bands of DM2 overlap with the omnidirectional reflection band of DM1 at any incident angle. Hence, if DM2 will be disposed on the top of DM1 (or vise versa), one can expect that the omnidirectional reflection band of such a heterostructure will start from the bottom of the first band of DM2 and end at the top of the second band of DM2.

To check it the numerical calculations of the reflection spectra through 10 layers of DM1, 10 layers of DM2 and 10 layers of DM1 on the top of 10 layers of DM2 are presented in figures 4.10, 4.11 and 4.12, respectively, at normal, 40 deg., and 80 deg. angles of incidence for both polarizations of incident light.

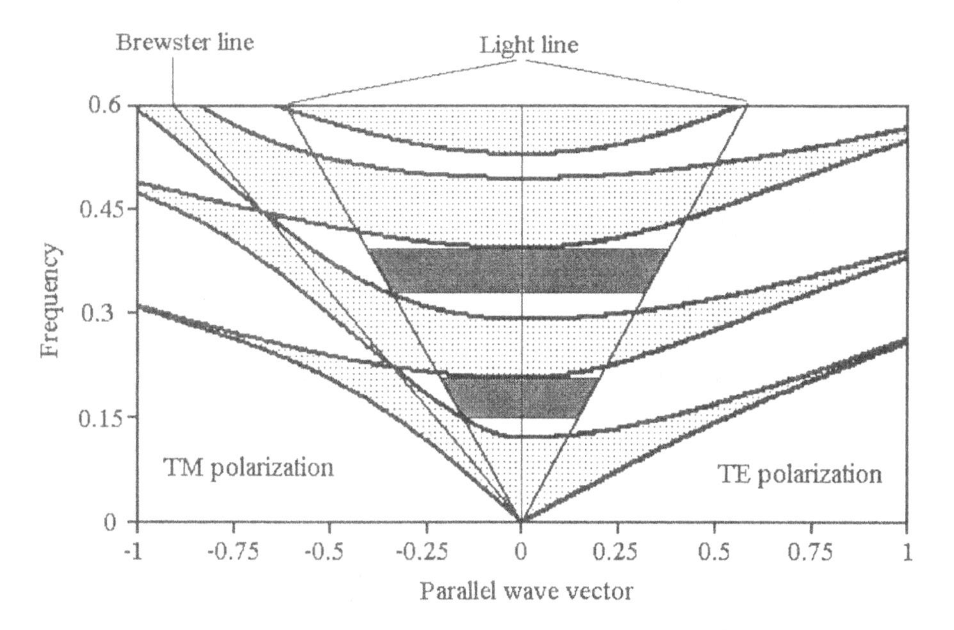

Figure 4.9. The numerically calculated band structure of the dielectric multilayer consisting of alternating layers of polystyrene and tellurium, with a thickness ratio of 0.55/0.45. The shaded zones are allowed bands; the parallel wavevector has the dimensions of $4\pi/\Lambda$, and frequency has the dimensions of $4\pi c/\Lambda$.

As follows from figures 4.10, 4.11 and 4.12, the omnidirectional reflection band of such a heterostructure is indeed enlarged with respect to the omnidirectional reflection bands of individual multilayers (DM1 and DM2) to over 80% of range to midrange, twice that of the DM1. Expansion to more than two multilayers in such a heterostructure to enlarge the omnidirectional frequency range of dielectric reflector is quite straightforward [47], although, in this case special attention should be directed toward the prevention of the Fabry-Perot-like valleys in the reflection spectrum. This can be done as with the classical reflectors case (see section 3.3.2 of this book) by placing the matching layer between the stacks.

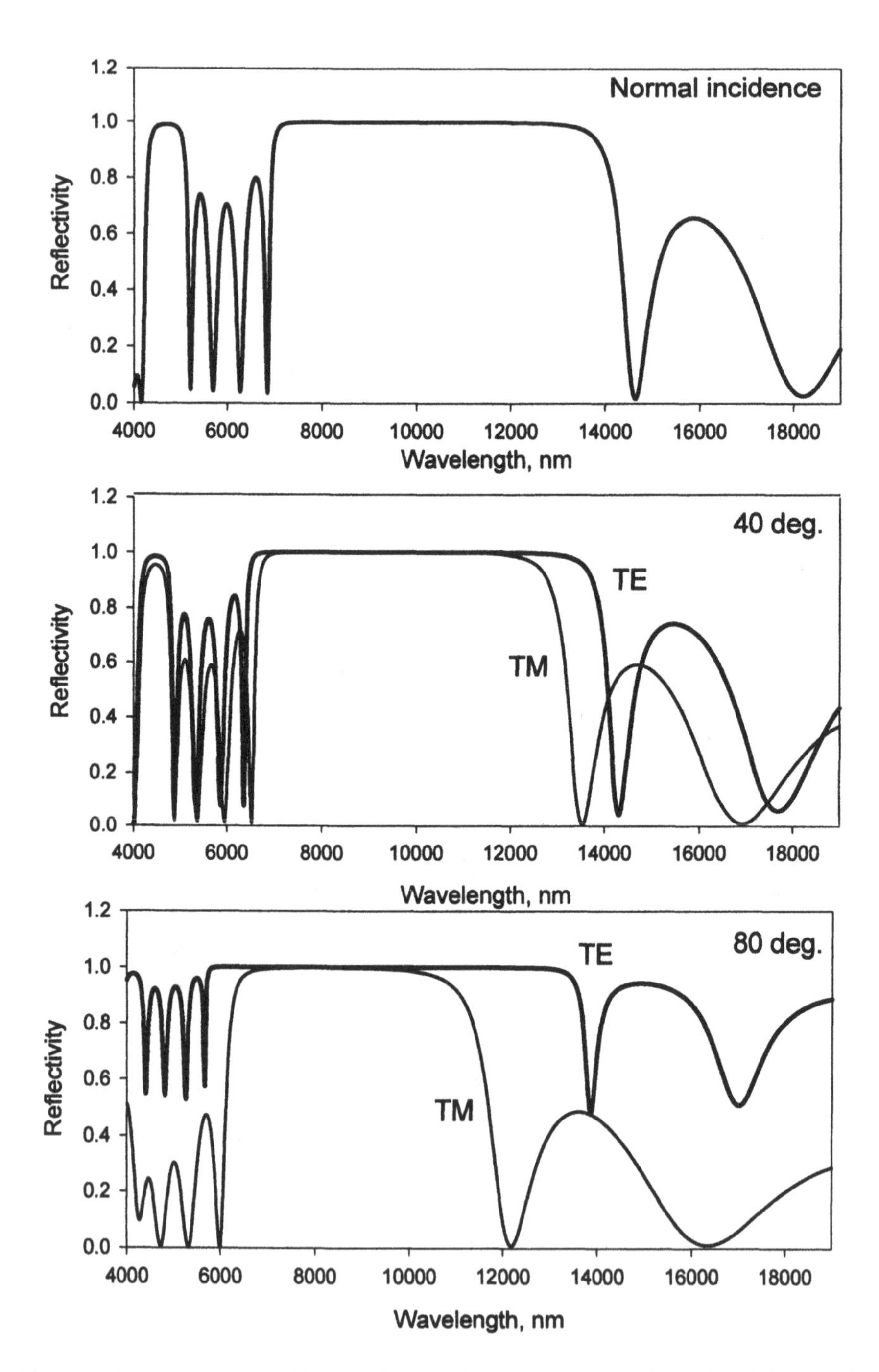

Figure 4.10. The numerically calculated reflectance spectra of a dielectric multilayer consisting of 10 alternating layers of polystyrene and tellurium, with a thickness ratio of 0.8/0.2 at normal, 40 deg., and 80 deg. angles of incidence.

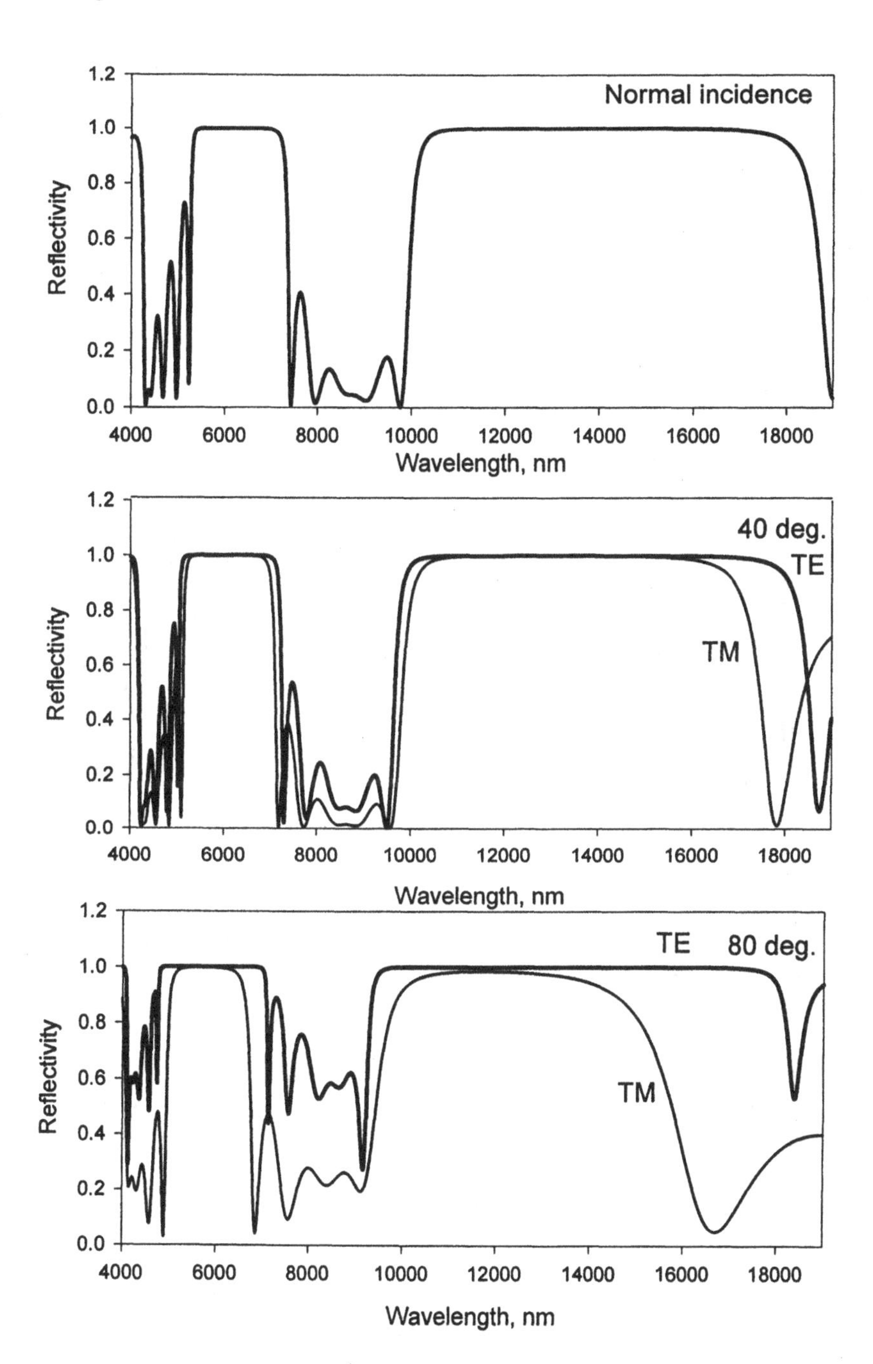

Figure 4.11. Numerically calculated reflectance spectra of a dielectric multilayer consisting of 10 alternating layers of polystyrene and tellurium, with a thickness ratio of 0.55/0.45 at normal, 40 deg. and 80 deg. angles of incidence.

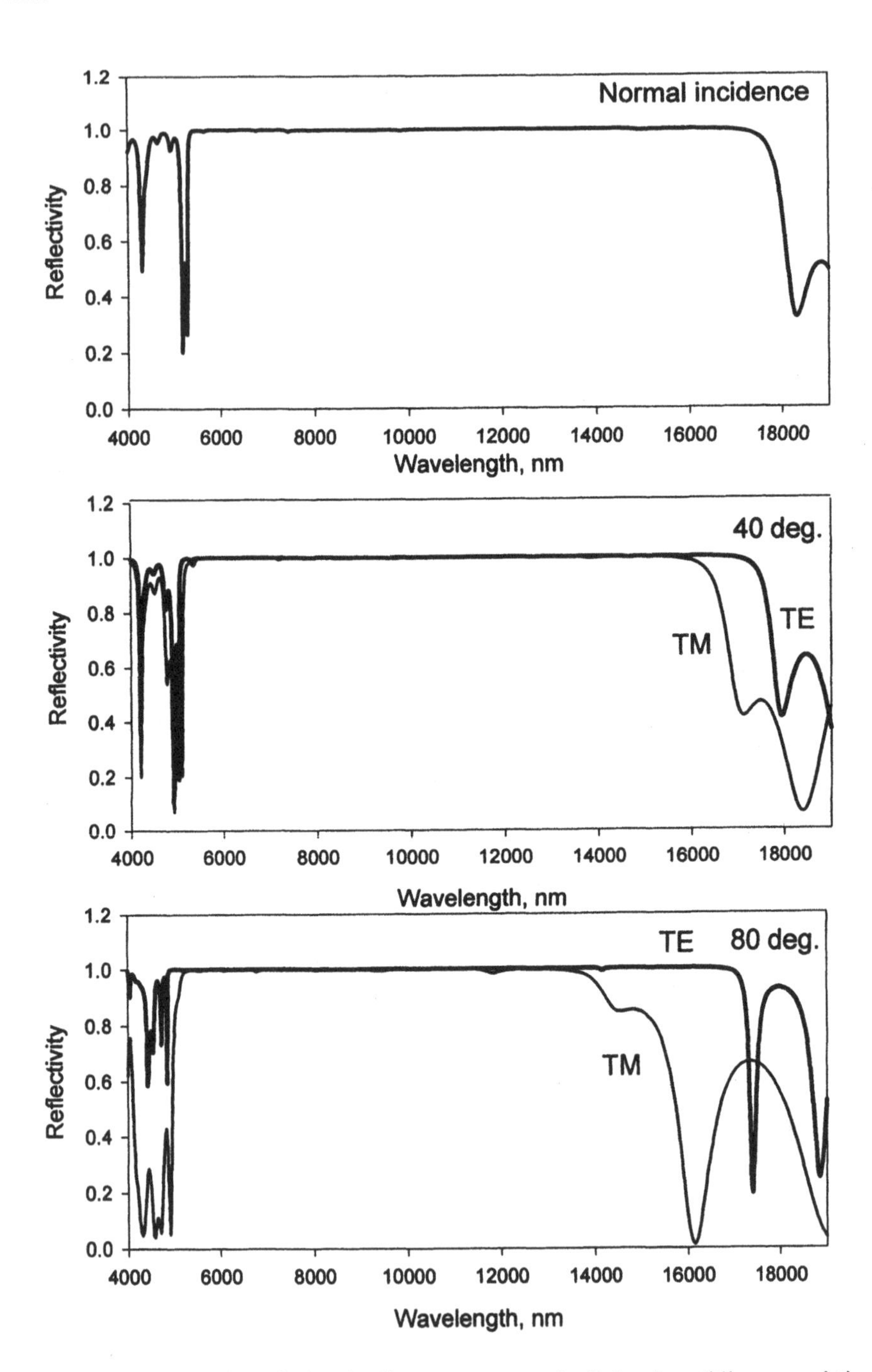

Figure 4.12. Numerically calculated reflectance spectra of a dielectric multilayer consisting of 10 alternating layers of polystyrene and tellurium, with a thickness ratio of 0.8/0.2 on the top of a dielectric multilayer consisting of 10 alternating layer of polystyrene and tellurium with a thickness ratio of 0.55/0.45 at normal, 40 deg., and 80deg. angles of incidence.

It should be noted that different approaches to enlarging the omnidirectional reflection band could be transferred from ordinary dielectric reflector designs (discussed in section 3.3.2 of this book). For example, one can change the periods through the multilayer to form a regular progression to ensure that, at any wavelength within a reasonably wide range, there are enough periods in the multilayer to give high enough omnidirectional reflection. This approach is especially important for the visible spectral range where there is no transparent high refractive index material (like tellurium of semiconductors in the near IR and IR spectral ranges) available.

Chapter 5

OMNIDIRECTIONAL SHORT-PASS FILTERS

5.1 Introduction

Many applications and instruments require the optical filters to be used in the ultraviolet, deep ultraviolet, or even far ultraviolet wavelength ranges. These applications include deep-UV focal-plane arrays for military applications, electrical spark imaging, water purification, blood chemistry analysis, and the chemical evaluation of foods, pollutants, and gases. As the wavelength of light becomes shorter in the ultraviolet range, existing optical filters (described in section 3.4 of this book) suffer from numerous disadvantages, such as poor optical performance, limited physical longevity, high autofluorescence, poor imaging quality of transmitted radiation, and transmitted wavelength instability.

No absorption-based filter material is known to have short-pass characteristics with transparency range down to deep or far UV. Dielectric film technologies for optical coatings employed for ultraviolet applications include deposition of soft, marginally adherent multilayer thin films onto various glasses. The films are soft and lack physical durability; most films are also water-soluble. Usually, such films consist of materials such as lead fluoride, cryolite (AlF_6Na_3), zinc sulfide, and so on. Coatings may also contain refractory metal oxides, which are generally more durable. However, standard oxide coatings are optically unstable when exposed to a varying environment. In addition, soft film filters can be sensitive to temperature and humidity and therefore have relatively limited operable lifetimes.

Different classes of ultraviolet optical filters employ thin films that are designated as *MDM* (*metal-dielectric-metal*), based on multiple-cavity metal-dielectric Fabry-Perot filter design, described in section 3.6.4 of this book. MDM filters are comprised essentially of a single substrate of fused silica or quartz, upon which a multilayer coating consisting of two materials, a dielectric (usually cryolite) and a metal (usually aluminum) is deposited. Unfortunately, MDM films are soft and easily damaged by moisture and oxygen. The final filter therefore consists of a second fused silica substrate mechanically fixed within a ring assembly with a vacuum or an atmosphere of argon separating the two substrates.

The MDM ultraviolet optical filter is generally a bandpass filter, which will pass a short range of wavelengths and eliminate out-of-band wavelengths by reflection. This type of filter is most commonly employed for deep UV applications (wavelengths shorter than 300 nm). Such filters are transmissive at a particular design wavelength band (pass band). MDM filters offer an advantage over soft-coating type filters by eliminating laminating epoxies, thus eliminating performance degradation due to solarization (UV discoloration). Unfortunately, as was discussed in section 3.6.4 of this book, the optical performance of MDM filters is rather limited. Typically, the peak transmission of 270 nm to 300 nm bandpass filters is at most about 10 to 25%. The maximum usable temperature of this filter type is relatively low, typically less than 150°C.

An additional disadvantage of all interference-based short-pass filters is the dependence of the transmission spectra on the angle of light incidence, as it was discussed in section 3.4.2 of this book.

In this chapter I present a new approach to short-pass filter design. It promises to provide filters that are as omnidirectional as absorption-based filters, are environmentally stable, have as good control over the short-pass edge position and steepness as interference-based filters, and have transmittance that considerably exceeds that of MDM-type filters.

5.2 Short-Pass Filters Based on Leaky-Waveguides Array

To understand the principle of operations of the short-pass filter of this chapter, we should remember the fundamental property of waveguides and leaky waveguides (see sections 2.6 and 2.8 of this book), the cut-off wavelength, and the behavior of the leaky waveguide loss coefficients near the mode cut-off. It means that for any waveguide structure such a wavelength exists below which such a structure will not support any waveguide mode (the same is true for leaky waveguides). Hence, an array of similar waveguides such as the transmission is permitted only through the waveguides modes, such a filter will exhibit a short-pass transmission spectrum shape, which will be independent of the angle of incidence within the angle's range, defined by waveguide numerical aperture.

Such short-pass filters were suggested for 100 μm and longer wavelengths (far IR spectral range) (examples can be found in [14], [15], and [16]). These short-pass filters were based on the array of metallic leaky waveguides (see figure 5.1). The manufacturing of such filters was based on drilling uniform holes in thin metal plate. By drilling such holes with a high density, the overall transmission of such filters can be made sufficiently high.

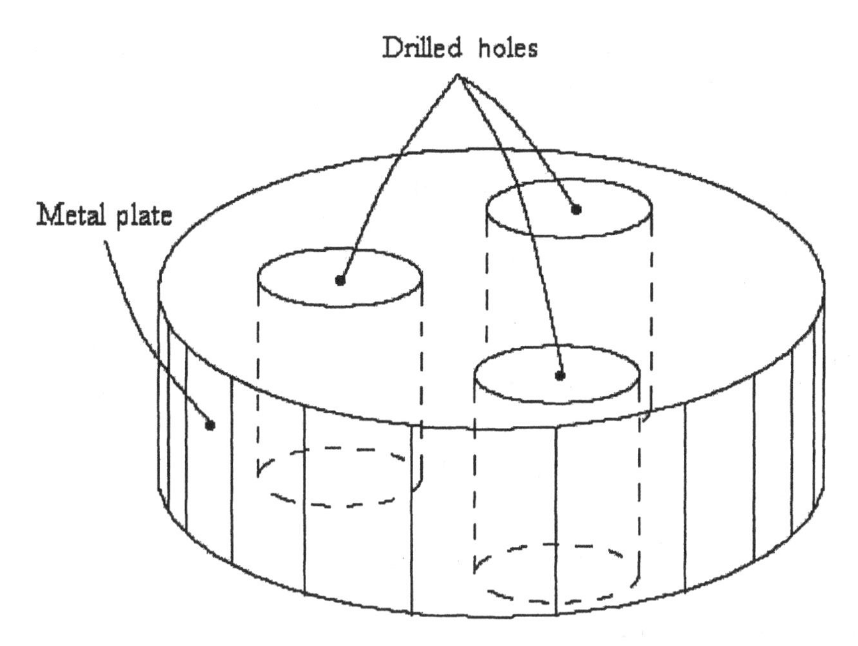

Figure 5.1. Waveguide mode cut-off IR short-pass filter design.

Metals are very good reflectors at the operational wavelengths of such filters, so the leaky waveguides propagation losses were low. These filters showed the following advantageous properties: they were rugged, generally consisting of a single piece of perforated metal or grid; they were relatively lightweight, compact, and not very sensitive to environmental factors such as heat and humidity. However, filter fabrication through hole drilling is a time-consuming process for far IR filters and is impossible for shorter wavelengths, since both the hole diameters and hole-to-hole spacing in such filters should be on the order of the designed short-pass edge. Even an IR wavelength of 2 microns will require holes with diameters of 2 microns. In addition, such filters have poor control over the rejection band edge steepness.

In [48], nanochannel glass with a follow-up covering of channel walls and both surfaces of the filter with highly reflective material, preferably metal, was recommended as a UV short-pass filter. However, such an approach does not give enough control over the shape of the transmission spectrum, particularly over the rejection band edge steepness. The transition from transmission to blocking of such filters in the UV range can take up to more than a 100 nm, which is unacceptable for most of the applications. Sharper transmission edge can be accomplished only with strong degradation of overall transmission efficiency. Another drawback includes the lack of control over the uniformity of channel sizes and channel wall smoothness, inherent to nanochannel glass.

A considerably more promising approach was suggested in [17]. In this paper free-standing macroporous silicon (MPSi) array was suggested as a material for short-pass filtering. The reasonable transmission was shown down to 180 nm wavelength, and it was predicted that the pass-band will be extended to far and even extreme UV (down to several tens of nanometers) (see figure 5.2). However, such short-pass filters have not provided truly short-pass behavior - the spectrum showed transmission in the near IR, which was unexpected [17].

In order to understand the performance of such filters (denoted in the future consideration as *Lehmann's filter*) better, let us consider the optical effects that take place during light transmission through the MPSi array. It should be noted that such principles are very similar to the metal waveguides filters, discussed above.

The structure of the Lehmann's filter is schematically shown in figure 5.3. Such a filter consists of air- or vacuum-filled macropores electro-etched into the silicon wafer host. The macropores are forming an ordered uniform array (in [17] the array was of cubic symmetry). The pore's ends are open on both the first and second interfaces of silicon wafer.

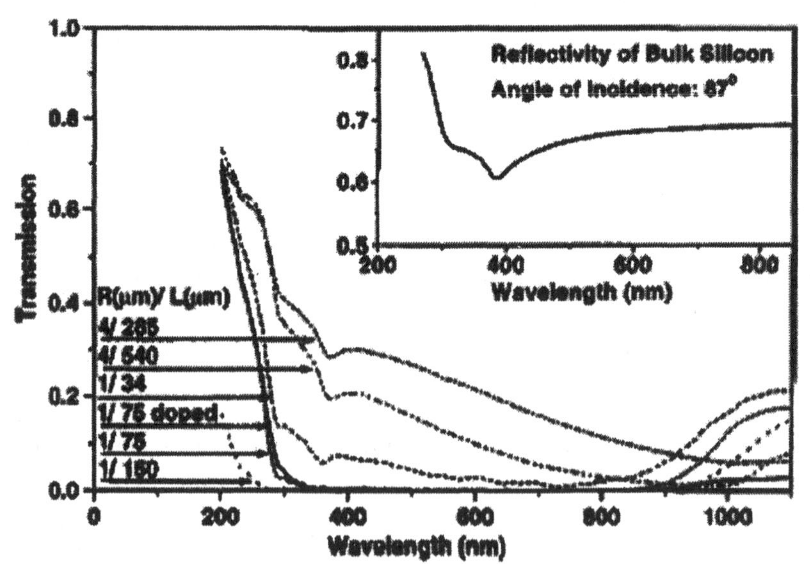

Figure 5.2. Measured transmittance (corrected for porosity) of MPSi array with pore radius R, pore length L, and the pore axis parallel to the light beam. Inset shows the spectral reflectivity of bulk silicon for a large angle of incidence (After [17]).

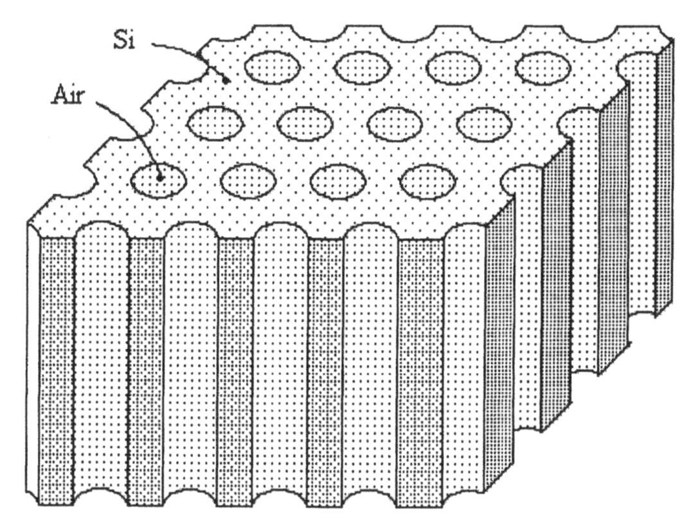

Figure 5.3. The short-pass MPSi filter structure of [17].

Silicon is opaque in the deep UV, UV, visible and part of the near IR wavelength ranges (see figure 5.4): the silicon absorption coefficient k is very high at the wavelengths below ~ 400 nm and high enough at wavelengths below ~ 900 nm to block all radiation coming through the silicon wafer with a thickness of 50 microns or more. Hence, light at wavelengths below ~700 nm can pass the structure shown in figure 5.3 only through the pores. Due to the comparability of the diameter of the pores (100 to 5000 nm) and the wavelength of light (200 to 1000 nm) and high aspect ratios in MPSi structures ($(t/d)>30$, where t is the MPSi layer thickness and d is the pore diameter), the transmission through MPSi layer structure at the wavelengths below 700 to 800 nm takes place through the leaky waveguide modes. The core of such leaky waveguides can be considered to be the air- or vacuum-filled pores, while the reflective walls (see the near metallic behavior of the refractive index n and absorption coefficient k of silicon given in figure 5.4 at wavelengths below ~370 nm) of such leaky waveguides are the pores walls being made of silicon. Hence, the MPSi layer can be considered as an array of leaky waveguides. Leaky waveguides can be considered as independent from each other and the visible, UV, and deep UV spectral ranges if they are separated by silicon with thickness more than 10 to 100 nm due to mentioned above high light absorption in silicon.

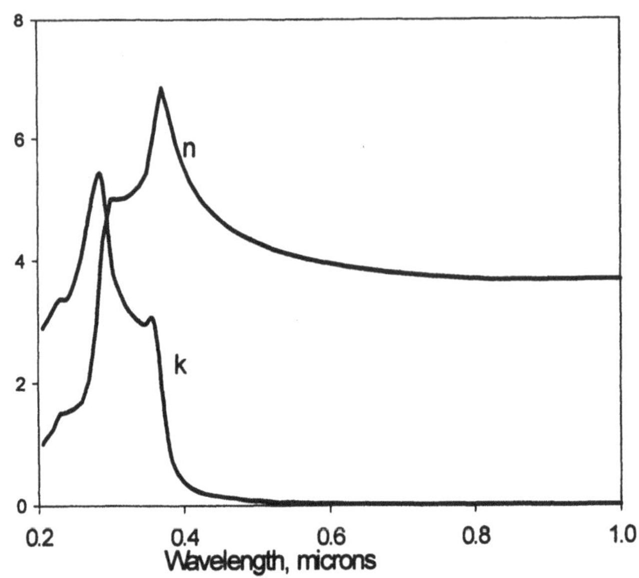

Figure 5.4. Spectral dependences of real and imaginary parts of silicon refractive index (after [39]).

In the near IR and IR wavelength range, the nature of the transmission through the MPSi layer of figure 5.3 changes. Silicon became less opaque at wavelengths of 700 to 900 nm and became transparent at wavelengths starting from approximately 1000 nm. The light at these wavelength ranges can pass through the MPSi layer not only through the pores but also through the silicon host (silicon "islands"). Due to the "islands" nature of the silicon host in the MPSi layer, the transmission at the wavelengths above ~700 nm will take place through waveguide modes confined in the silicon "islands." As a high refractive index material, silicon can support waveguide modes if surrounded by a lower refractive index material (air or vacuum in the MPSi filter of figure 5.3). Close packing of the pores is essential for efficient transmission through the MPSi layer in the UV and deep UV ranges. Hence, the MPSi layer can be considered in the near IR and IR ranges as an array of silicon waveguides in an air host.

The natural parameter to characterize the spectral dependence of the transmission through the MPSi filter will be the optical loss coefficient, α, having dimensions cm^{-1} (see the discussion in section 2.8 of this book). The amount of light still remaining in the pore leaky waveguide (or silicon waveguide) after it travels a length l is proportional to $\exp(-\alpha(\lambda)\, l)$, and the light remaining in the MPSi array at a distance l from the first MPSi interface is equal to

$$I(l,\lambda) = I_0\, P(\lambda) \exp(-\alpha(\lambda)\, l) \tag{5.1}$$

where I_0 is the initial intensity of the light entering the pore and $P(\lambda)$ is the coupling efficiency on first MPSi interface, which is discussed in more detail below. The optical loss coefficient is a function of pore size, geometry, distribution, and wavelength. It also depends on the smoothness of the pore walls. Roughness of walls introduces another source of absorption of light, scattering, which is in turn dependent on the wavelength to the mean square roughness ratio.

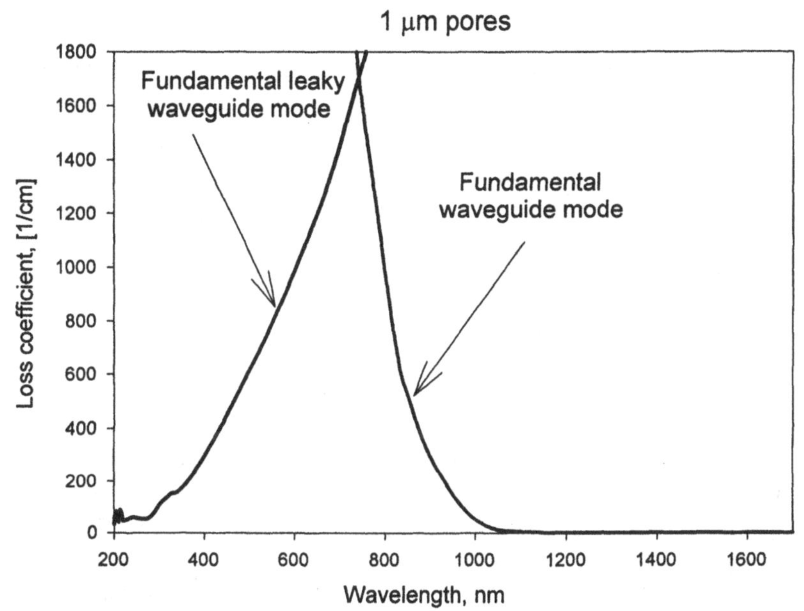

Figure 5.5. Numerically calculated spectral dependences of the optical fundamental leaky waveguide and waveguide modes loss coefficients for 1x1 μm^2 near-square pore in silicon.

Figure 5.5 shows numerically calculated (using the methods discussed in sections 2.6 to 2.8 of this book) spectral dependences of the optical fundamental leaky waveguide and waveguide modes loss coefficients for 1x1 μm^2 near-square vacuum-filled pores in silicon. According to figure 5.5, transmission through the pores leaky waveguides is dominant up to ~700 nm, the transmission through the silicon host waveguides is dominant starting from ~800 nm, while at 700 to 800 nm both transmission mechanisms compete with each other. The increase of the losses through leaky waveguides with the wavelength is due both to the reduction of the reflection coefficient of silicon and the redistribution of the leaky waveguide mode over the pore cross section (the modal field penetration into the silicon host material as well as the optical loss increases with the wavelength of light). Even with the assumption of zero dispersion material, the optical loss for the fundamental mode propagating in a leaky waveguide is proportional to the wavelength squared.

Figure 5.6 demonstrates numerically calculated dependences of effective refractive indices (a) and loss coefficients (b) of TE leaky waveguide modes on the pore cross-section for the near-square pores at the wavelength of 250 nm. The loss coefficients are decreasing with the increase of the pore size due to the mode intensity redistribution inside the pore described above. As follows from figure 5.6, the pore leaky waveguides became multimode at deep UV wavelength starting from pore size of approximately 220 nm. For example, for the 1-μm pore, the number of TE polarized modes will be eight for 250 nm wavelength.

As was shown above, both leaky waveguides at the deep UV, UV, and visible spectral ranges and waveguides at near IR and IR spectral ranges can be either single mode (supporting only the fundamental mode) or multimode (higher-order modes are also supported) depending on pore sizes and geometry. Hence, (5.1) should be generalized, and, in the general case, the amount of light remaining at distance l from the first filter surface will be

$$I(l,\lambda) = I_0 \Sigma P_i^{LW}(\lambda) \exp(-\alpha_i^{LW}(\lambda)\, l) + I_0 \Sigma P_i^{W}(\lambda) \exp(-\alpha_i^{W}(\lambda)\, l) \qquad (5.2)$$

where i is the number of the mode, numbered as follows: $i = 0$ corresponds to the fundamental mode; $i = 1$ corresponds to the first-order mode and so on; $P_i^{LW}(\lambda)$ is the coupling efficiency into i^{th} leaky waveguide mode; $P_i^{W}(\lambda)$ is the coupling efficiency into i^{th} waveguide mode; $\alpha_i^{LW}(\lambda)$ is the loss coefficient of ith leaky waveguide mode; and $\alpha_i^{W}(\lambda)$ is the loss coefficient of ith waveguide mode. The summation should be done through all modes supported by the given MPSi structure. For the leaky waveguides the losses increase rapidly with the increasing of i. It means that in general only the one or two leaky waveguide modes are responsible for the transmission at the deep UV, UV, and visible spectral ranges through MPSi structure with reasonable thickness (> 20 microns) for the pore diameters up to 1.5 microns, while all light coupled into higher-order leaky waveguide modes will be absorbed while traveling through the leaky waveguides. The coupling efficiency is the highest for the fundamental mode and in general decreases with the number of mode increases.

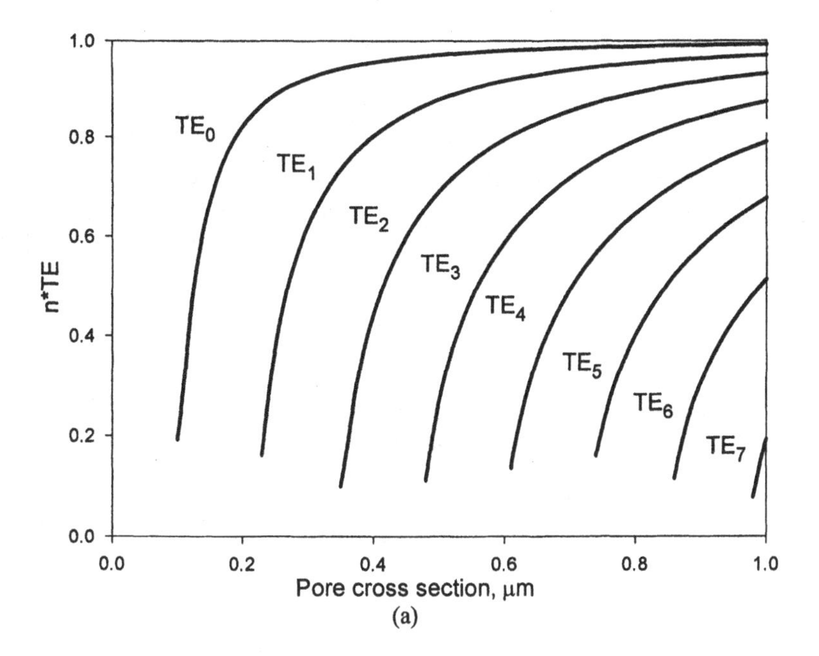

(a)

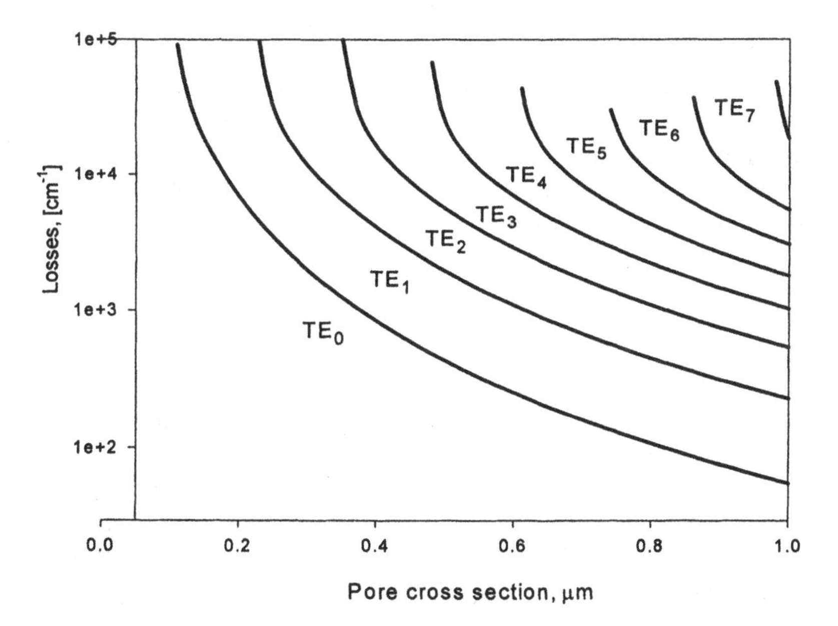

(b)

Figure 5.6. The numerically calculated dependences of effective refractive indices (a) and loss coefficients (b) of TE polarization leaky waveguide modes (a) and on the pore size for the leaky waveguides with near-square pores at the wavelength of 250 nm.

There are other parameters managing overall MPSi layer transmission. These parameters are the coupling efficiency of incident light into the leaky waveguide or waveguide modes at the first MPSi layer interface and outcoupling from the leaky waveguide or waveguide modes to transmitted light at the second MPSi layer interface. The efficiency, $P_i^W(\lambda)$, for the waveguide modes or $P_i^{LW}(\lambda)$ for the leaky waveguide modes of the coupling of normally incident light at the first MPSi layer interface into the MPSi structure is calculated as

$$P_i^W(\lambda) = \frac{\int E_i^W(s,\lambda)\cdot E_I^*(s,\lambda)\cdot ds}{\sqrt{\int E_i^W(s,\lambda)\cdot (E_i^W)^*(s,\lambda)\cdot ds \cdot \int E_I(s,\lambda)\cdot E_I^*(s,\lambda)\cdot ds}} \quad (5.3a)$$

$$P_i^{LW}(\lambda) = \frac{\int E_i^{LW}(s,\lambda)\cdot E_I^*(s,\lambda)\cdot ds}{\sqrt{\int E_i^{LW}(s,\lambda)\cdot (E_i^{LW})^*(s,\lambda)\cdot ds \cdot \int E_I(s,\lambda)\cdot E_I^*(s,\lambda)\cdot ds}} \quad (5.3b)$$

where E_i^W (s,λ) is the electric field of an ith waveguide mode at the plane of the first MPSi layer interface; E_i^{LW} (s,λ) is the electric field of a ith leaky waveguide mode at the plane of the first MPSi layer interface; E_I(s, λ) is the electric field of incident wave on the MPSi layer perpendicular to the mode propagation direction; and λ is a wavelength of light. If a plane-parallel beam of light is normally incident on the first MPSi layer interface, for the fundamental leaky waveguide mode the (5.3a) can be simplified to

$$P_0^{LW}(\lambda) \approx \frac{S_p}{S_{uc}} \quad (5.4a)$$

where S_p is the area of each pore (see figure 5.3), which for the circular pore cross section is equal to $\pi\, \o^2/4$, where $\o$ is the diameter of the pore and for near-square pore cross section is ~ d^2, where d is the characteristic cross section size; and S_{uc} is the area of an MPSi array's unit cell (which could be introduced for ordered MPSi arrays only). For the waveguide transmission (for near IR or IR wavelength ranges) (5.3b) can be also simplified to

$$P_0^W(\lambda) \approx \frac{4n_{Si}(\lambda)\cdot n_I}{(n_{Si}(\lambda)+n_I)^2}\cdot\frac{S_{uc}-S_p}{S_{uc}} \quad (5.4b)$$

where $n_{Si}(\lambda)$ is the refractive index of the silicon at the wavelength λ and n_I is the refractive index of the incident medium. For the most common case of air, (5.4b) can be rewritten as

$$P_0^W(\lambda) \approx \frac{4n_{Si}(\lambda)}{(n_{Si}(\lambda)+1)^2} \cdot \frac{S_{uc}-S_p}{S_{uc}} \tag{5.5}$$

In other words, to some approximation, $P_0^{LW}(\lambda) \approx p$ and $P_0^W(\lambda) \approx 0.69(1\text{-}p)$, where p is the porosity of MPSi layer. It should be noted that the approximation given above for waveguide case (for near IR and IR wavelength ranges) is not as good as for leaky waveguide case (deep UV, UV, and visible spectral ranges) due to strong cross-coupling between neighbor waveguides.

At the second surface of the MPSi layer, the light from each end of each leaky waveguide or waveguide is emitted with a divergence governed by the numerical aperture of the leaky waveguide or waveguide, *NA*, and wavelength. In the far field, the destructive and constructive interference of all light sources in the form of leaky waveguides (or waveguides) ends will take place. In the case of an ordered MPSi array, this leads to a number of diffraction orders, and this number is in general defined by the pore array geometry (by the relationship between pore size and pore-to-pore distance) and the wavelength of light. As an example, the angular distribution of diffraction efficiency for a 2 μm period cubic MPSi array with 1.5 x 1.5 μm near-square pores at 275 nm wavelength is presented in figure 5.7.For most optical filters applications, only the 0th diffraction order, the efficiency of which is denoted in the following discussion as $DE_0^{LW}(\lambda)$ for a leaky waveguide array and $DE_0^W(\lambda)$ for the waveguide array, is of interest. The only application known to the author that is not sensitive to the outcoupling of light to higher diffraction orders is that of the filter mounted directly on top of photodetector. In all other cases, the outcoupling losses at the second MPSi layer interface can be estimated as 1- $DE_0^{LW}(\lambda)$ for a leaky waveguide array or as 1 - $DE_0^W(\lambda)$ for the waveguide array, since the main source of such losses is the redistribution of light into higher diffraction orders. Such losses are sensitive on both wavelength and pore array geometry. They are more pronounced at short wavelengths due to the higher number of diffraction orders. These losses are higher for hexagonal arrays than for square arrays, but due to the higher density of leaky waveguides or waveguides, the total transmission efficiency through hexagonal MPSi arrays is generally higher.

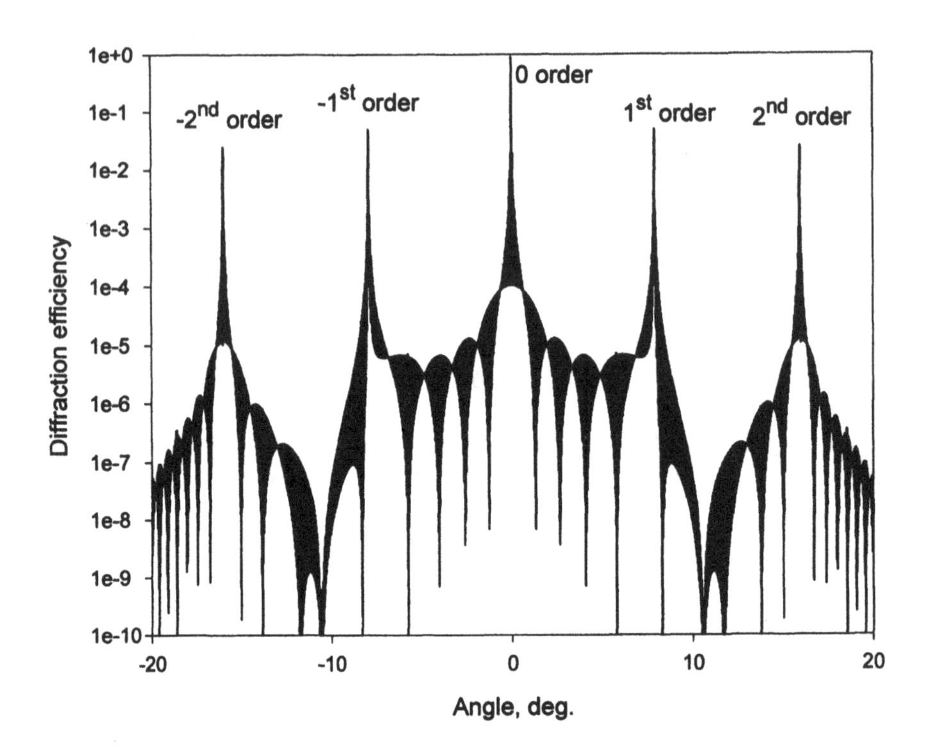

Figure 5.7. The angular distribution of diffraction efficiency for a 2 μm period cubic MPSi array with 1.5x1.5 μm near-square pores at 275 nm wavelength.

The overall transmission spectra $T(\lambda)$ through MPSi layer with the thickness L (where L is high enough that the transmission through higher-order modes could be neglected) can be estimated at normal incidence of light as

$$T(\lambda) \approx I_0 \{ pDE_0^{LW}(\lambda) \exp(-\alpha_0^{LW}(\lambda) L) + 0.69(1-p) DE_0^{W}(\lambda) \exp(-\alpha_0^{W}(\lambda) L)\} \qquad (5.6)$$

It should be noted that outcoupling losses can be completely suppressed for any given wavelength if the MPSi array period is on the order of this wavelength. For instance, a 280 nm wavelength in the "solar-blind" region of the spectrum that is important for many applications will require a pore array period on the order of 280 nm and pore diameters of about 100 nm. Manufacturing such a filter is possible in principle and expected to become cost-effective.

In figure 5.8 the plot of the spectral dependence of transmission through an MPSi array of cubic symmetry, 50 μm thickness and pore, and 1 μm pore diameters is given, numerically calculated according to (5.6). Close matching with experimental data of [17], given in figure 5.2, validates the model discussed above.

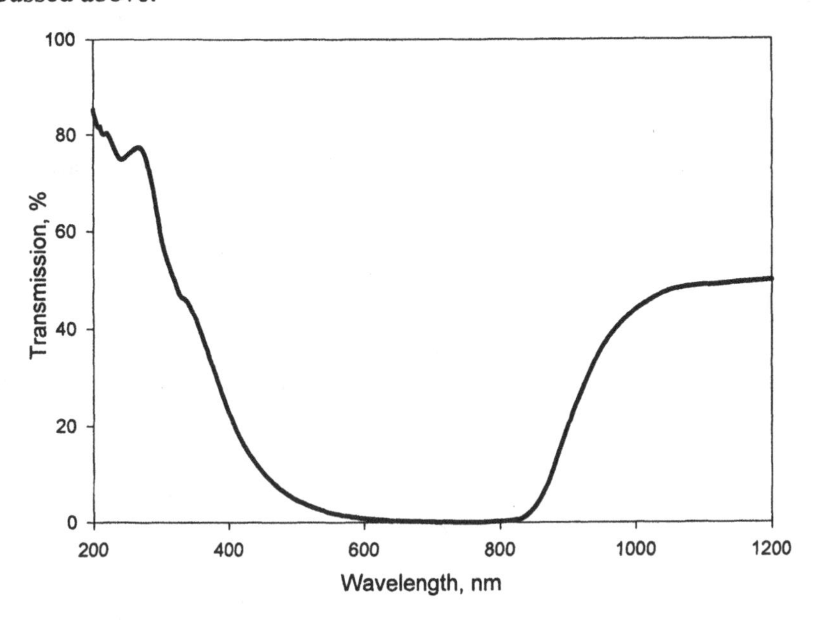

Figure 5.8. The numerically calculated spectral dependence of transmission through MPSi array of cubic symmetry, 50 μm thickness, and 1 μm pore diameters.

As was mentioned previously in the text, MPSi-based short-pass filter design has considerable advantages with respect to existing deep and far UV filters. These include transparency down to the far UV spectral range, omnidirectionality of the transmittance spectra shape within the acceptance angles of the leaky waveguides (which will be discussed in more detail later in this book), and other features listed above. However, such a short-pass filter design does not provide enough control over the steepness and the position of the rejection zone edge. As follows from figures 5.2 and 5.8, the transition from rejection to transmission of such a filter takes 100 nm or more, which is not acceptable for most applications. As follows from equation (5.6) and figures 5.5 and 5.6, it is possible to increase the steepness of the edge by increasing the thickness of MPSi layer or by decreasing the pore sizes, which, however, will cause a necessary shift of the edge position to shorter wavelengths. Still, it is unrealistic for present manufacturing capabilities to achieve MPSi array parameters (pores cross sections 500 nm or more); to get the edge width to 30 nm one will need (see equation (5.6) and figures 5.5 and 5.6) the MPSi layer thickness must be at least 200 μm thick (i.e., aspect ratio of 400). Unfortunately, MPSi layers with such parameters have not yet been obtained.

Fortunately, the transmission spectra of MPSi layer can be managed by managing the α_0^{LW} term in (5.6). It can be done by covering the pore walls with a transparent dielectric multilayer, that is, by creating a multilayer leaky waveguide array.

5.3 Multilayer Leaky Waveguide Short-Pass Filter Design

The simplest realization of a multilayer leaky waveguide array is a free-standing MPSi layer with pore walls uniformly coated by a single layer of transparent dielectric material, as is shown in figure 5.9. Such a layer will strongly modify the spectral dependences of leaky waveguide loss coefficients by means of constructive and destructive interference of the leaky waveguide mode inside this layer.

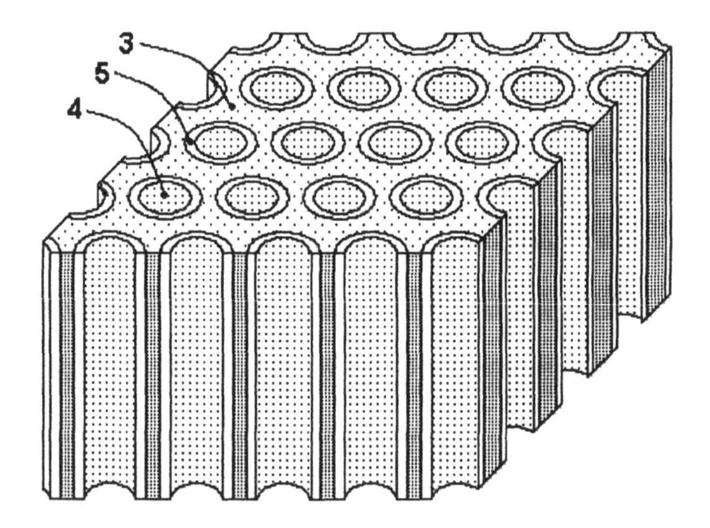

Figure 5.9. Freestanding macroporous silicon uniform pores array with one layer of transparent dielectric material uniformly covering pores walls.

The diagrammatic view of such a structure is given in figure 5.9. Such an MPSi filter will consist of silicon host 3, macropores 4 and the layer of dielectric transparent material 5 uniformly covering pore walls. Layer 5 can be, for example silicon dioxide (SiO_2), thermally grown or deposited through the *low-pressure chemical vapor deposition* (*LP CVD*) technique. The fabrication of such structures will be discussed in more detail in section 5.5 of this book.

It should be noted that the material of layer 5 does not have to be transparent over the transmission range, rejection range, and/or rejection edge. Depending on the particular filter design, this material can be sufficiently transparent through rejection range and/or rejection edge, while being absorptive or reflective at most parts of the transmission range. The optical thickness of layer 5 can be arranged so it serves as an antireflective layer for the leaky waveguide modes for the wavelengths inside the rejection range near the initial (pure silicon) rejection edge of MPSi filter to make the rejection edge of the final optical filter sharper and to increase the suppression inside the blocking range of such optical filter. For example, silicon dioxide layer can serve as an antireflection layer for pores leaky waveguides starting from 170 nm wavelength and above (SiO_2 is transparent at these wavelengths) while for shorter wavelengths SiO_2 will reflect light the same as silicon.

As an example, in figure 5.10 the numerically calculated spectral dependences of loss coefficients (a) and transmittance through the 50 μm layer (b) of fundamental TE modes are presented for pure silicon MPSi layer and MPSi layer with pore walls covered by 70 nm of silicon dioxide. Pores were assumed to have a near-square cross section of 1 μm. The silicon dioxide layer thickness was chosen so that the silicon dioxide layer serves as an antireflection layer for the wavelength in the middle of the rejection edge of initial (uncoated) MPSi filter (~300 nm). Figure 5.11 shows numerically calculated spectral dependences of reflectivity from a pure silicon surface and silicon coated by 70 nm of silicon dioxide at 85 deg. of incidence (corresponded to the wavevector of pore leaky waveguide mode around 300 nm wave). The suppression of the reflection of the leaky waveguide mode from the pore walls (see figure 5.11), causes a high and relatively narrow peak of mode losses centering around the wavelength of reflectivity minimum (see figure 5.10a). The rejection level of MPSi the layer with pores, covered by silicon dioxide, exceeds that of the pure silicon MPSi layer, while the steepness of the slope of losses coefficient peak is much sharper. The losses at transmission range of silicon dioxide-covered MPSi layer are only slightly higher than those for the pure silicon MPSi layer.

Figure 5.10b illustrates the sharpening of the rejection edge by at least five times for silicon dioxide-coated MPSi layer with respect to the uncoated MPSi layer. The decrease of the transmission inside the pass band is only 5% less than that of the uncoated MPSi layer.

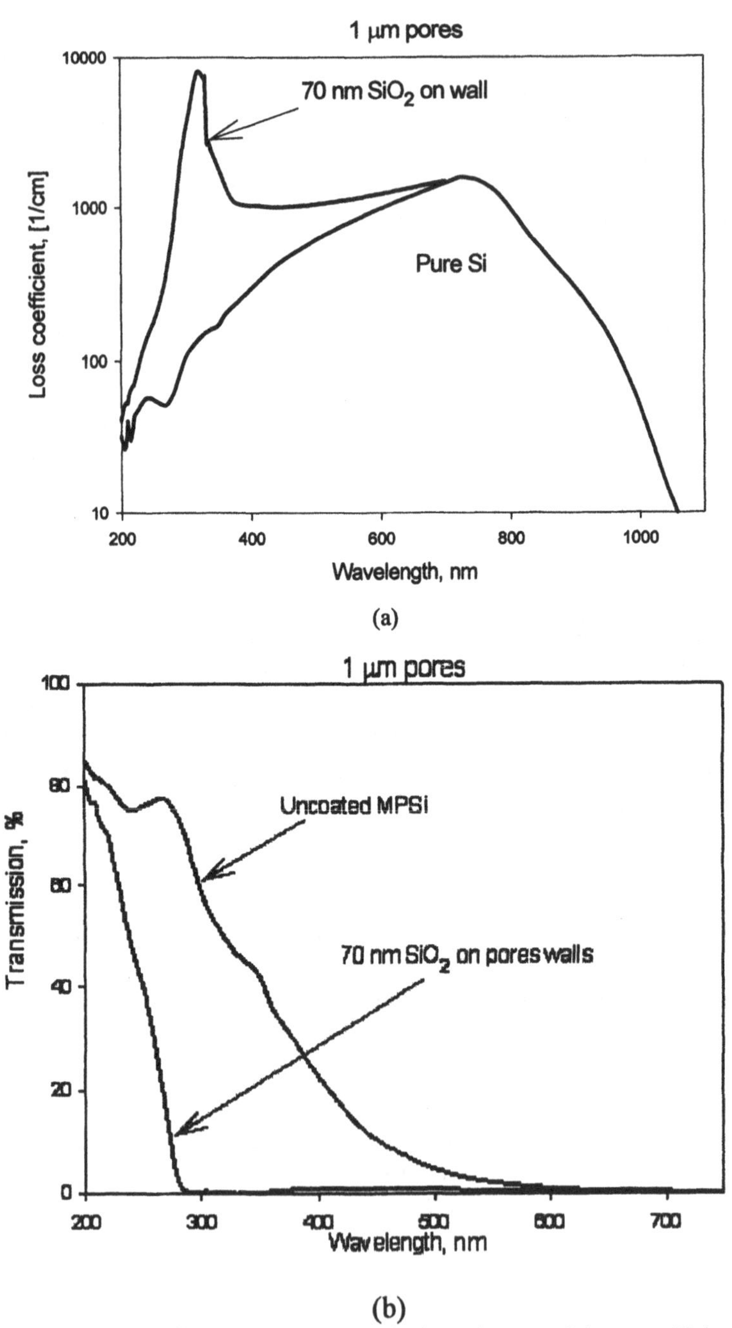

Figure 5.10. Numerically calculated spectral dependences of loss coefficients (a) and transmittance through 50 μm layer (b) of fundamental TE modes for pure silicon MPSi layer and MPSi layer with pores walls covered by 70 nm of silicon dioxide. Pores were assumed to have a near-square cross section of 1 μm.

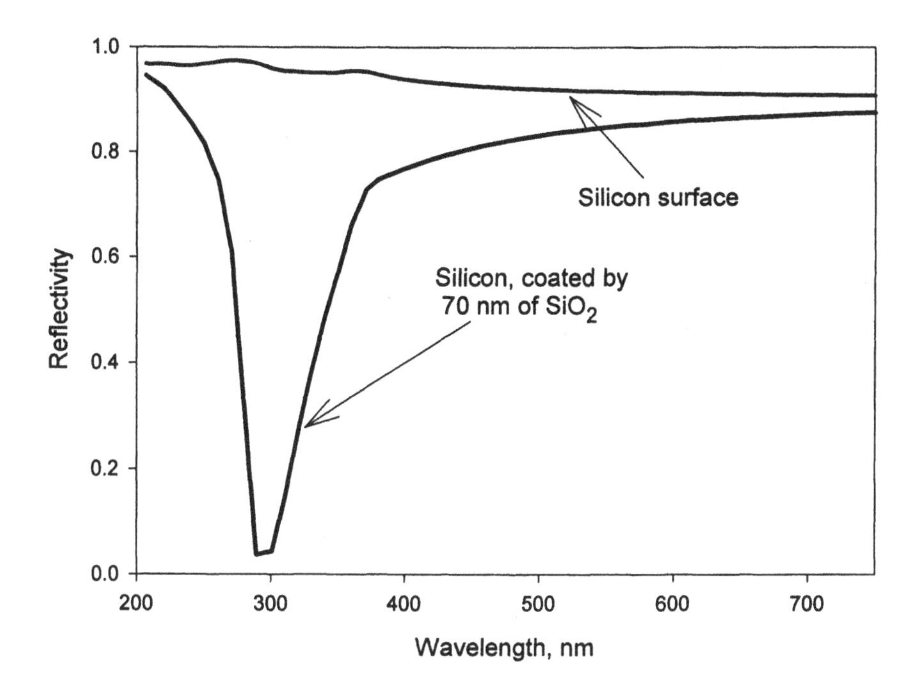

Figure 5.11. Numerically calculated spectral dependences of reflectivity from a silicon surface and silicon coated by 70 nm of silicon dioxide at 85 deg. of incidence.

The same as for common single-layer antireflection coating (see section 3.1.1 of this book), it is possible for a coated MPSi layer to tune the wavelength position of the rejection edge by varying the antireflection layer thickness. As an example, in figure 5.12 the numerically calculated spectral dependences of loss coefficients (a) and of transmittance through the 50 μm layer (b) of fundamental TE modes are given for an uncoated MPSi layer and MPSi layer with pores walls covered by 40, 70, and 100 nm of silicon dioxide. Pores were assumed to have near-square cross section of 1 μm.

As expected, the rejection edge of spectral filters can be tuned in the wavelength from below 200 nm for 40 nm silicon dioxide thickness to about 300 nm for 100 nm silicon dioxide thickness while keeping the rejection edge much sharper than that of the uncoated MPSi layer and the transmittance with the pass band on about the same level.

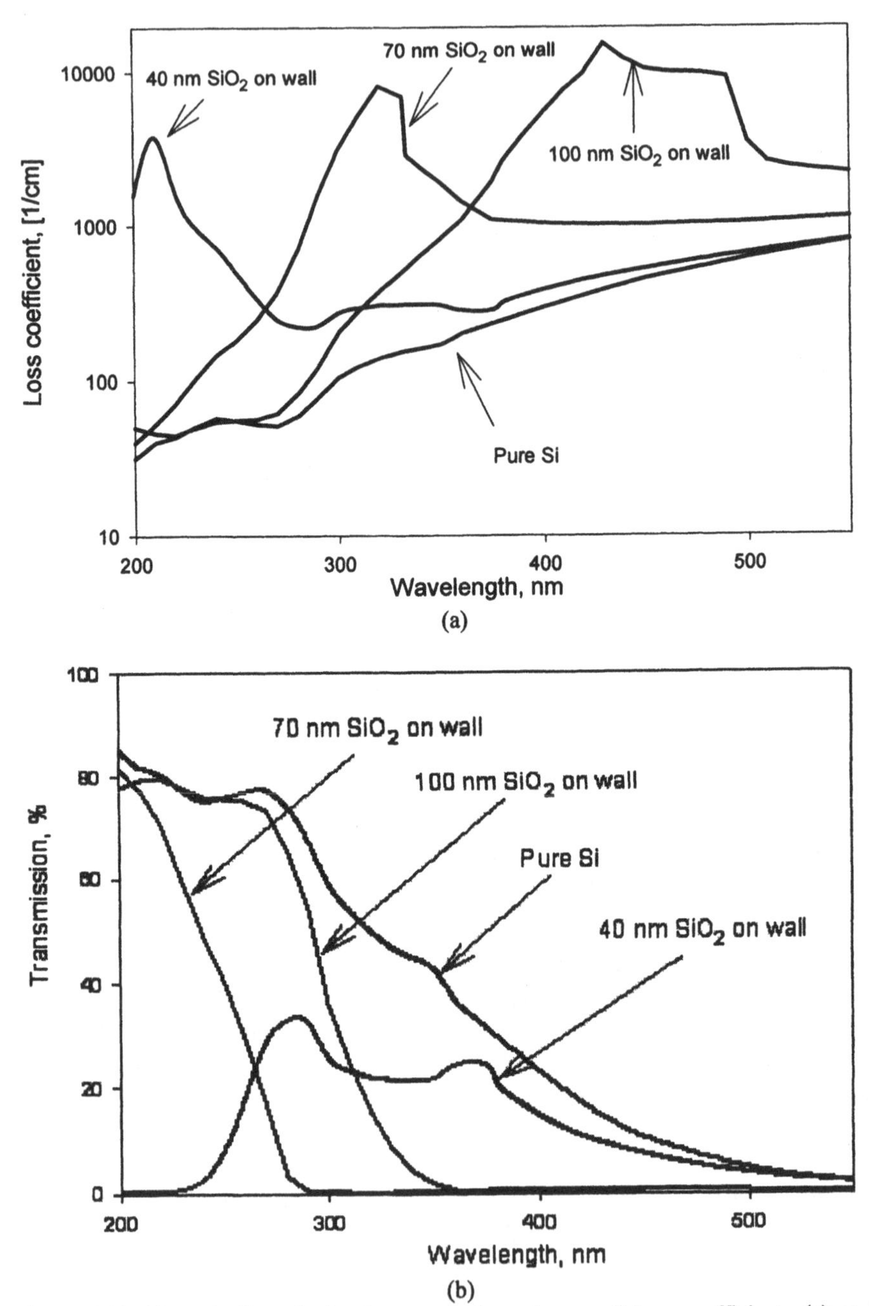

Figure 5.12. Numerically calculated spectral dependences of loss coefficients (a) and of transmittance through the 50 μm layer (b) of fundamental TE modes for the pure silicon MPSi layer and the MPSi layer with pore walls covered by 40, 70, and 100 nm of silicon dioxide. Pores were assumed to have a near-square cross section of 1 μm.

Figure 5.13 shows numerically calculated spectral dependences of reflectivity from a pure silicon surface and silicon coated by 40, 70, and 100 nm of silicon dioxide at 85 deg. of incidence. The peaks of loss coefficients from figure 5.12a are quite close to the minimums of reflection in figure 5.13. However, the coincidence is not full, and the loss coefficient spectral dependences are considerably more complicated than the reflectivity from the plane interface. It happens due to variations of propagation constant of the leaky waveguide modes in the pores with the wavelength (see figure 5.6a) or, in other words, due to different reflection angles for the different wavelengths and due to nonplanar character of the electromagnetic wave in the leaky waveguide mode (see figure 2.16 from chapter 2 of this book).

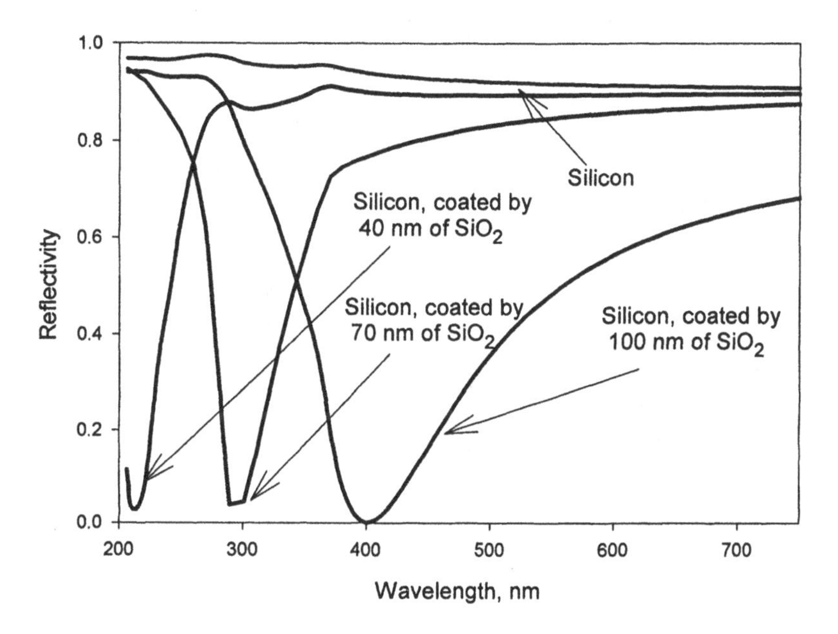

Figure 5.13. Numerically calculated spectral dependences of reflectivity from a pure silicon surface and silicon coated by 40, 70, and 100 nm of silicon dioxide at 85 deg. of incidence.

As follows from figure 5.12.b, the presence of the silicon dioxide layer on the pore walls influences not only the position of the rejection edge but also the rejection inside the rejection zone. For a 40 nm silicon dioxide layer on the pore walls, the peaks of the transmission appear in the middle of the rejection zone (at about 280 and 370 nm, respectively). These peaks of transmission correspond to the deeps in loss coefficient spectral dependence (figure 5.12a) and peaks of reflectance in figure 5.13. By comparing it with the real and imaginary parts of the silicon refractive index spectra (figure 5.4), we can conclude that these unwanted peaks are caused by the peculiarities of silicon refractive index wavelength dependence. Unfortunately, with single-layer pore wall coating it is impossible to control the position of the rejection edge and the rejection through the whole rejection zone of MPSi layer simultaneously.

The short-pass filter made of silicon dioxide-coated free-standing MPSi layer has some commercial perspectives as a short-pass filter. However, just as with single-layer antireflection coatings (see section 3.1 of this book), the suppression of the reflection of the leaky waveguide mode from the pore walls is spectrally limited to a relatively narrow spectral range. This causes the appearance of unwanted transmission peaks inside the filter rejection zone and limited freedom in short-pass edge position and shape. Similar to antireflection coatings (see section 3.12), by adding more layers of different materials on the leaky waveguides (i.e., pore) walls, it is possible to gain considerably better control over the MPSi filter performance.

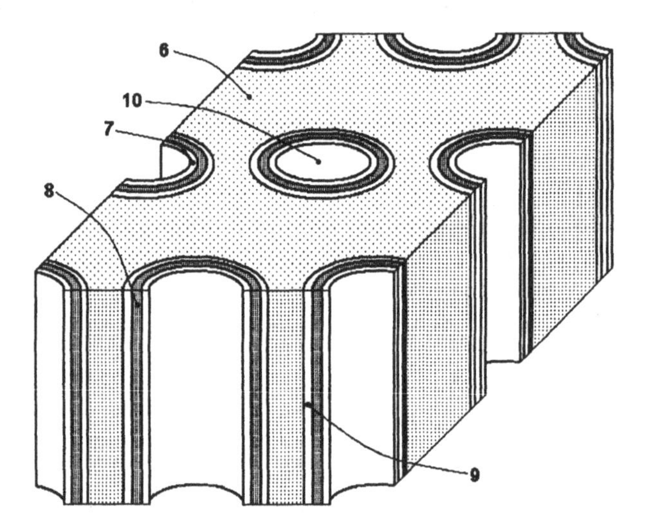

Figure 5.14. Freestanding macroporous silicon uniform pore array with multiple layers of transparent dielectric materials uniformly covering pores.

The diagrammatic view of an MPSi layer with pore walls covered by several layers of (at least two) different transparent dielectric materials is given in figure 5.14. Such a structure consists of a silicon host (6), macropores (10), and layers of transparent dielectric materials (7, 8, 9) uniformly covering pore walls. A three-layer coating is shown as an example. The number of layers in a multilayer coating can be considerably more and will be determined by particular design and application requirements (i.e., by a tradeoff between performance and price). Just as for the single-layer coating, the materials composing a multilayer coating do not have to be transparent over the pass band, rejection band, and rejection edge of such a filter. Hence, silicon nitride, silicon dioxide, magnesium-, calcium-, and barium fluorides, cryolite, and other materials transparent at UV range can be used for deep and far UV filter.

As it is for a single layer of dielectric material covering pore walls, the main purpose of having multilayer cover pore walls is to modify the spectral dependence of leaky waveguide modes loss coefficient ($-\alpha_0^{LW}(\lambda)$). However, unlike a single-layer coating, which can be arranged only as an antireflection layer on the top of high refractive index silicon substrate, the multilayer coating can be arranged as a high-dielectric reflector inside the pass band or as a wide-band antireflection coating inside the rejection band of such an MPSi filter.

Figure 5.15a gives the numerically calculated spectral dependences of loss coefficients for fundamental TE modes of an uncoated MPSi layer and an MPSi layer with pores walls covered with 70 nm of silicon dioxide and a five-layer coating. Pores were assumed to have a near-square cross-section of 1 μm. The structure of the five-layer coating was as follows: {silicon pore wall/ 48 nm of silicon dioxide/23 nm of titanium dioxide/ 59 nm of silicon dioxide/33 nm of silicon nitride/ 116 nm of silicon dioxide/ air (inside the pore)}. The multilayer coating was designed so that the multilayer serves as a dielectric mirror for the wavelength of 250 nm (i.e., within the MPSi layer pass band) and an angle of incidence of 85 deg. It is clearly illustrated by figure 5.15a that with five-layer dielectric coating on the pore walls, the leaky waveguide mode losses within the pass band of the MPSi layer are suppressed by an order of magnitude with respect to uncoated MPSi layer and by more than order of magnitude comparing to the single-layer antireflection coating discussed above. At the same time the leaky waveguide mode losses inside the rejection band of an MPSi layer coated with five-layer high-reflectance coating exceed those of an uncoated MPSi layer by an order of magnitude and are at the same level as that of the single-layer coated MPSi. The short wavelength slope of loss coefficient peak is much sharper than that of both uncoated and single-layer coated MPSi. These advantageous features allow considerably thicker layers of MPSi to be used, which makes rejection within the rejection band of the MPSi layer considerably higher, while the rejection edge is considerably sharper than that of uncoated and single-layer coated MPSi.

Figure 5.15b demonstrates the numerically calculated spectral dependences of transmittance through the 200 μm layer for fundamental TE modes of an uncoated MPSi layer and an MPSi layer with the same five-layer coating that was discussed above. One can see that for an MPSi layer with five-layer coated pore walls the transmission edge is sharper by at least 10 times over an uncoated MPSi layer. Moreover, the transmission efficiency inside the pass band of such an MPSi layer is about two times higher than the transmission efficiency of an uncoated MPSi layer having the same thickness.

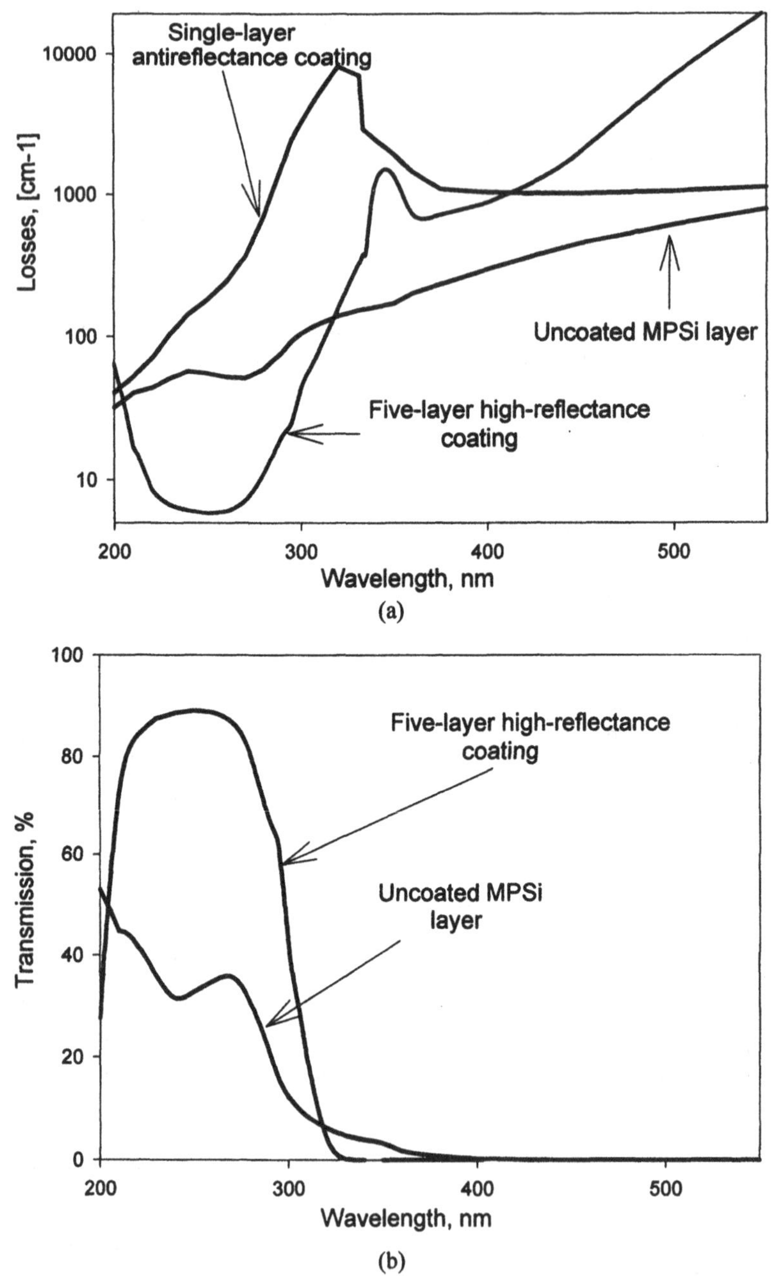

Figure 5.15. Numerically calculated spectral dependences of losses coefficients (a) and transmittance through 200 μm layer (b) of fundamental TE modes for uncoated MPSi layer and MPSi layer with pore walls covered by 70 nm of silicon dioxide and five-layer high-reflectance at 250 nm coating. Pores were assumed to have a near-square cross-section of 1 μm.

The rejection edge width of ~ 20 nm makes such a layer a potentially valuable commercial filter. It should be noted that the edge could be made even sharper by increasing the thickness of such an MPSi layer. However, it will cause some scarification of transmittance within the pass band. As is illustrated by figure 5.15b, the width of the pass band for a five-layer coated MPSi layer is limited, unlike the case of the uncoated MPSi layer, which, as was mentioned above, is expected to be transmittive down to far and extreme UV wavelength ranges. It is caused by the limited width of the high-reflectance band of multilayer coating (see section 3.3.2 of this book). It is possible to increase the width of the high-reflectance band of the coating and, through that, of the pass-band of the MPSi layer by increasing the number of layers in multilayer or by increasing the refractive index contrast within the multilayer. This approach, however, is rather limited due to the lack of the transparent in the deep UV materials (for this realization of pore wall coating, all layers within the multilayer have to be transparent within the pass band of the MPSi layer).

The important feature of an MPSi layer with pores coated by the transparent dielectric multilayer is the tremendous value of rejection within the rejection band and the width of the rejection band. Figure 5.16 gives the numerically calculated transmission of the same MPSi layers as figure 5.15b in the logarithmic scale. According to this plot, for the five-layer coated MPSi the rejection within the rejection band (from 300 to ~750 nm) exceeds 10 orders of magnitude and exceeds 20 orders of magnitude for most of the rejection band.

Such values of the rejection over the visible spectral range is clearly higher than of any interference-based deep-UV short pass filters (as was mentioned in the beginning of this chapter no other types of deep UV short-pass filters are available). The transmittance of such a filter in the near IR and IR spectral ranges, however, will more or less repeat that of an uncoated MPSi layer (see figures 5.2 and 5.8). Such an MPSi layer will be more or less transparent at these wavelength ranges.

Recalling equation (5.6) and the discussion above, the transmission at near IR and IR spectral ranges takes place through waveguide modes confined inside the intrapore silicon islands:

$$T\,|_{\text{Near IR, IR}} = \Sigma\, DE_i^{W}(\lambda)\, P_i^{W}(\lambda)\, \exp\left(-\alpha_i^{W}(\lambda)\, l\right) \qquad (5.7)$$

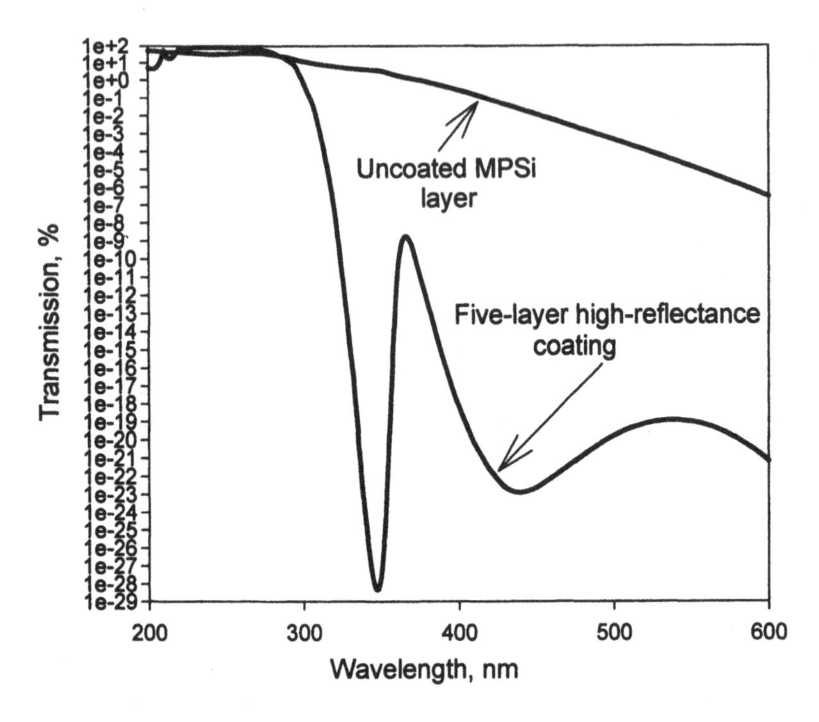

Figure 5.16. Numerically calculated spectral dependences of transmittance (shown in logarithmic scale) through the 200 μm layer of fundamental TE modes for the uncoated MPSi layer and MPSi layer with pore walls covered by five-layer high-reflectance at 250 nm wavelength coating. Pores were assumed to have near-square crossection of 1 μm.

It is possible to suppress the transmission though waveguide modes (5.7) completely by covering first, second, or both surfaces of the MPSi layer by thin (100 to 200 nm) metal film. In this case the coupling efficiency of light to waveguides will be completely suppressed: $P_i^W(\lambda) = 0$. By doing that one can obtain a truly short-pass deep UV filter: rejection of all wavelengths longer than the rejection edge would be tremendous (10 orders of magnitude or more).

It should be noted, however, that the covering of MPSi layer surfaces with a metal layer will sacrifice both coupling and outcoupling efficiencies. This will also somewhat suppress the coupling efficiencies. It will be caused by decreasing the open area of each leaky waveguide by the area of the multilayer coating cross-section. Fortunately, most of the leaky waveguide mode energy is concentrated within the air (or vacuum) core of the leaky waveguide. Figure 5.17 illustrates the numerically calculated fundamental mode intensity and electric field distribution across the pore cross-section for the fundamental TE-polarized leaky waveguide mode for the MPSi layer with pore walls covered by the five-layer coating at the wavelength of 250 nm (central wavelength of the pass-band of MPSi layer).

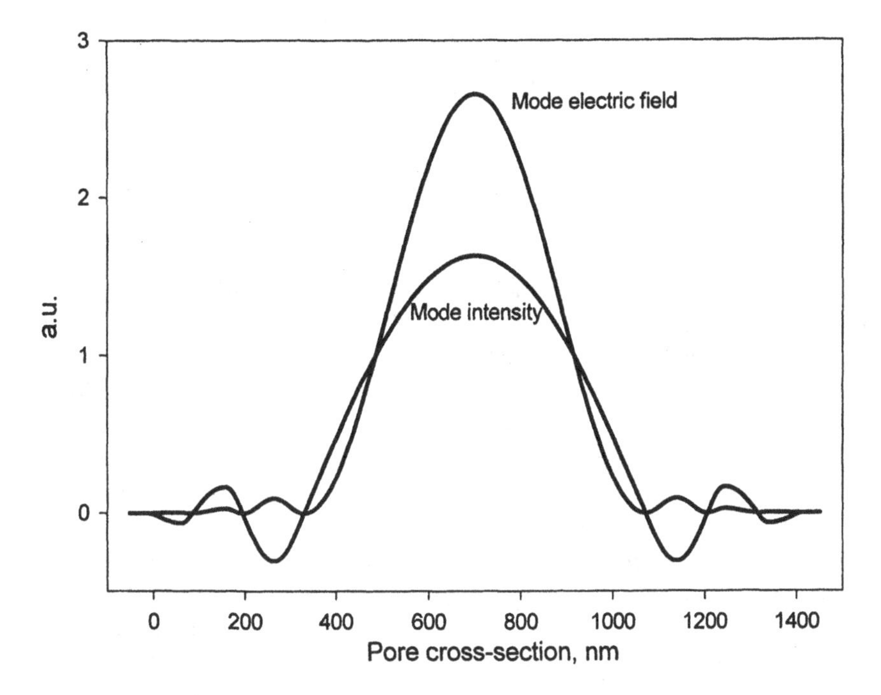

Figure 5.17. Numerically calculated fundamental mode intensity and electric field distribution across the pore cross-section for fundamental TE modes for the MPSi layer with pore walls covered by a five-layer high-reflectance at 250 nm coating. Pores were assumed to have near-square cross-section. The wavelength was assumed to be 250 nm.

It is possible for the coated MPSi layer to tune the wavelength position of the pass band by varying the structure of multilayer coating, just as for multilayer dielectric high-reflectance coatings (see section 3.3.2 of this book). As an example, figure 5.18a presents the numerically calculated spectral dependences of loss coefficients of the fundamental TE mode for an uncoated MPSi layer and for an MPSi layers with two different five-layer dielectric coatings. Pores were assumed to have a near-square cross-section of 1 μm. The structures of multilayer coatings were as follows: first coating: {silicon pore wall/ 48 nm of silicon dioxide/23 nm of titanium dioxide/ 59 nm of silicon dioxide/33 nm of silicon nitride/ 116 nm of silicon dioxide/ air (inside the pore)}; second coating: {silicon pore wall/ 62 nm of silicon dioxide/28 nm of titanium dioxide/ 71 nm of silicon dioxide/ 38 nm of silicon nitride/ 141 nm of silicon dioxide/ air (inside the pore)}. As expected, the leaky waveguide mode loss minimum of the MPSi layer with a second coating is shifted toward the longer wavelength with respect to the leaky waveguide mode loss minimum of MPSi layer with first coating. However, the tuning of the pass-band position of the coated MPSi layer is a considerably more complex task than the tuning of the short-edge of ordinary multilayer filter. Figure 5.18a illustrates that the shifting the position of the leaky waveguide mode propagation loss minimum is accomplished by reshaping of the spectral dependence of whole mode propagation losses. To avoid it, one needs to take into account not only the structure of the multilayer coating on the pore walls but also the diameter (for the round pores) or cross-section (for the near-square pores) of the pores in the MPSi array.

Figure 5.18b presents the numerically calculated spectral dependences of loss coefficients of fundamental TE mode for an uncoated MPSi layer and for MPSi layers with the same as in figure 5.18a five-layer dielectric coatings. As expected, the pass band of the MPSi layer with the second coating is shifted toward the longer wavelength with respect to the pass-band of MPSi layer with first coating, just as with the leaky waveguide mode propagation loss minimums. The already mentioned above reshaping of the mode losses spectral dependence did cause the reshaping of the transmission shape of the second coating with respect to the first coating. In particular, the long wavelength rejection edge of the MPSi layer with the second coating became two times wider than that of the MPSi layer with the first coating. However, the transmission efficiency within the pass band and the rejection level within the rejection band are on about the same level for both MPSi layers with first and second coatings. Taking into account the arguments given during the discussion of figure 5.18a, we can conclude that it is possible to design short-pass MPSi filters that have a high level of performance with different spectral positions of the pass bands throughout deep UV and UV wavelength ranges.

As mentioned above, the multilayer coating of the pore walls can be arranged not only as a high reflector for the wavelength within the pass band of MPSi layer. A multilayer structure can be chosen, for example, so the coating suppresses the reflection within the rejection band of the MPSi layer. In other words, it will serve as a multilayer antireflection coating for the rejection band spectral range. Unlike the single-layer antireflection coating discussed above, the multilayer antireflection coating can be arranged so the reflection is suppressed within a considerably wide spectral range (see the discussion in section 3.1.2 of this book). From another point of view, it is possible to create such a multilayer coating on the walls of the MPSi layer that the reflection will be suppressed within the rejection band and enhanced within the pass band of MPSi layer, similar to the multilayer edge filter discussed in section 3.4.2 of this book.

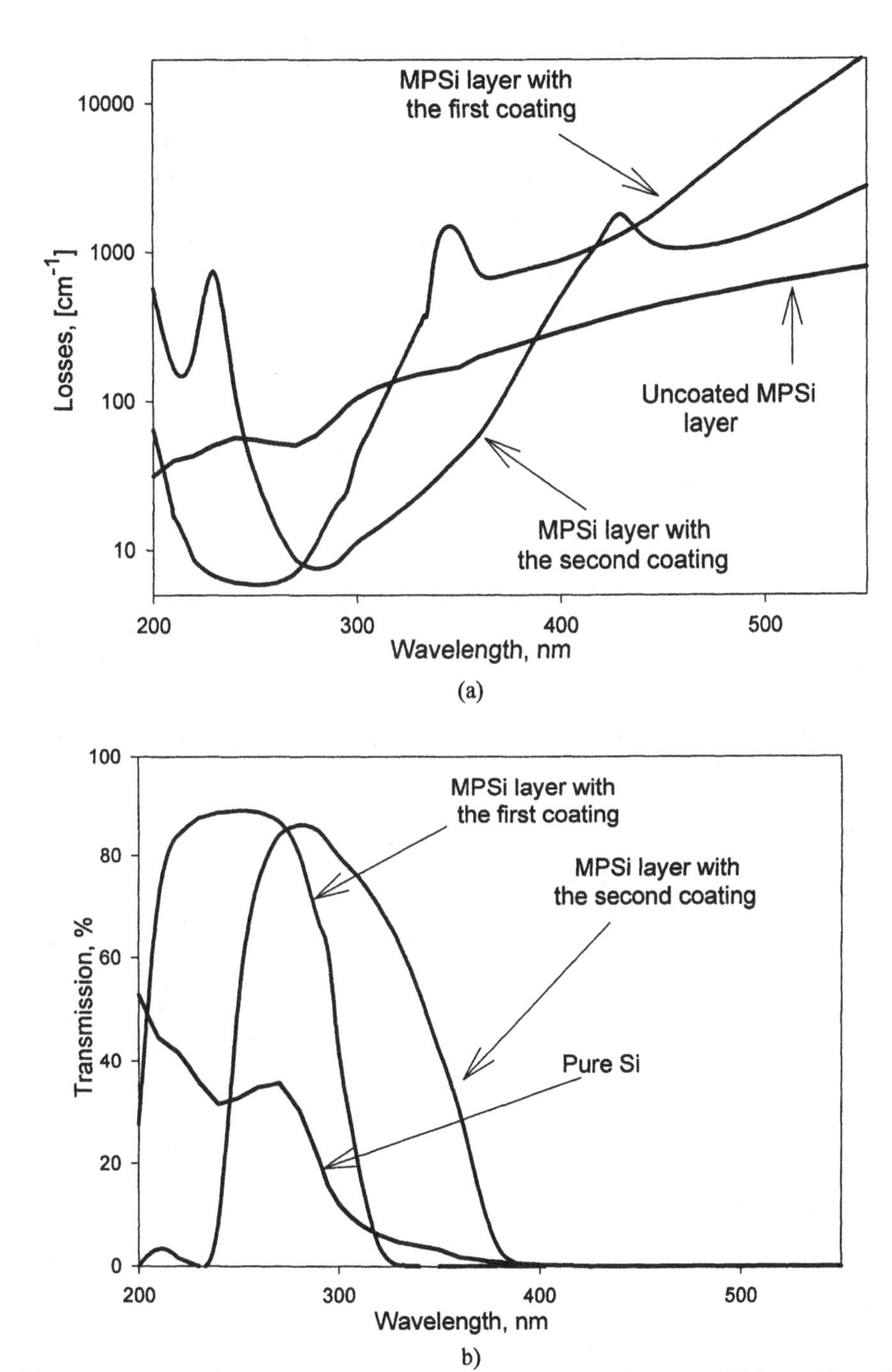

Figure 5.18 Numerically calculated spectral dependences of loss coefficients (a) and of transmittance through the 200 μm layer (b) of fundamental TE modes for the uncoated MPSi layer and the MPSi layer with pore walls covered by different five-layer dielectric coatings.

Such a multilayer coating of the MPSi layer pore walls will provide the following advantages over ordinary short-pass filters: the steepness of the rejection edge will be increased (due to multiple reflections that the light-waves experience during propagation through leaky waveguide), the rejection level will be higher by many orders of magnitude, and the transmission spectrum will be independent of the angle of incidence.

Although the mechanism of improving the transmission spectra of the MPSi layer with multilayer-coated pore walls is interference-based, the MPSi layer will not suffer from the disadvantages of the usual interference filters, such as the dependence of the spectrum on angle of incidence (see discussions in chapter 3 of this book). Let us neglect the transmission through the silicon waveguides. We can do that since the losses of waveguide modes are high in the short wavelength part of the spectrum (see figure 5.5) from one point of view and since this transmission can be completely suppressed by the coating either surface of the MPSi layer with metal, as was discussed above. In this case we can write

$$T(\lambda,\theta,\theta')\,|_{\text{deep UV, UV, VIS}} = \\ = \Sigma\, P_i^{LW}(\lambda,\theta)\, DE_i^{LW}(\lambda,\theta')\, \exp(-\alpha_i^{LW}(\lambda)\, l) \qquad (5.8)$$

where θ is the angle of incidence of light on the first MPSi layer surface and θ' is the angle created by the light reemitted from the second surface of the MPSi layer and the second MPSi surface. As was mentioned above, the $\exp(-\alpha_i^{LW}(\lambda)\, l)$ part of (5.8) contributes most to the overall transmission spectral shape of the MPSi layer. $\exp(-\alpha_i^{LW}(\lambda)\, l)$ is independent for each leaky waveguide mode on angle of incidence, since the mode loss mechanism has nothing to do with light coupling to or outcoupling from the leaky waveguide. However, there are other parameters that we need to take into account. For thin MPSi layers (with thickness below 20-50 μm, depending on pore cross-section), we cannot neglect the multimode nature of pore leaky waveguides. In this case the relative coupling efficiencies into different order modes will be angularly dependent, and some angular dependence of the transmission spectral shape of the MPSi layer is expected. However, as discussed above, for good performance of the MPSi layer as a filter, the thickness of the MPSi layer has to be at least 100 to 200 μm. It follows from figure 5.6b, that for such a thickness of the MPSi layer, the transmittance through all but fundamental leaky waveguide modes can be neglected. In this case (5.8) can be rewritten as follows:

$$T(\lambda,\theta,\theta')\,|_{\text{deep UV, UV, VIS}} = \\ = P_0^{LW}(\lambda,\theta)\, DE_0^{LW}(\lambda,\theta')\, \exp(-\alpha_0^{LW}(\lambda)\, l) \qquad (5.9)$$

The coupling efficiency at the first MPSi layer surface can be estimated as

$$P_0^{LW} = p \exp(-(\theta / \theta_{ac})^2) \tag{5.10}$$

where p is the porosity of the MPSi layer and θ_{ac} is the acceptance angle of the pore leaky waveguide. As discussed in Chapter 2 of this book, the particular value of the acceptance angle strongly depends on the leaky waveguide structure. For silicon pore leaky waveguides, the θ_{ac} is around 8 to 14 degrees depending on pore wall coating. Since we are investigating uniform ordered MPSi arrays and are interested in 0-order diffraction (see discussion above), $\theta' \equiv \theta$ and the dependence of diffraction efficiency is expected to be $DE_0^{LW}(\lambda, \theta) = DE_0^{LW}(\lambda, \theta = 0) \exp(-(\theta / \theta_{ac})^2)$. Hence, the dependence of the transmission on the angle of incidence of the coated MPSi layer will be closer to absorption-based filters (Schott glass filters, colored glass filters) and will gradually decrease when the angle of incidence goes away from the normal direction within the acceptance angle of the leaky waveguide mode, while the shape of the transmission spectrum will not change.

Figure 5.19a shows the numerically calculated transmittance spectra through the interference short-pass filter (having a structure similar to that shown in figure 3.18a from section 3.5 of this book) for normal incident, 10- and 20-deg. tilted plane-parallel beams. The wavelength shift of the pass-band edge position common to all interference edge filters is demonstrated. Figure 5.19b presents the normalized transmittance (as was discussed above, maximum transmittance will be defined by the particular filter structure and can vary from ~ 40% to 75%) spectra through an MPSi layer with a five-layer coating (the same as in figure 5.15) for normal incident, 20- and 30-deg. tilted plane-parallel beams. As follows from figures 5.19a and 5.19b, a coated MPSi layers should provide considerable advantages over interference edge filter designs and will provide the opportunity to use short-pass filters at different angles of incidence ($\pm$20 deg. at least). Figure 5.19c illustrates the numerically calculated transmittance spectra through the interference short-pass filter of figure 5.19a for normal incident beams with different convergence: plane-parallel beam (0-covergence angle) and Gaussian beams with 20 and 40 deg. convergence angle. The degradation of both the band-edge shape and out-of-band rejection, common to the interference short-pass filter, are demonstrated. Figure 5.19d presents the normalized transmittance spectra through an MPSi layer with 5-layer coating for 0, 20 and 40 deg. convergent normally incident Gaussian beams. It follows from figures 5.19c and 5.19d, that the MPSi layer with pore walls coated by a dielectric multilayer will provide the opportunity to use multiple-cavity narrowband pass filters at convergent or divergent beams ($\pm$40 deg. at least). Taking into account the extremely high rejection level and the infinite width of the rejection band (all wavelengths longer than the short-pass edge), filters based on the MPSi layer if developed to the level sufficient for commercialization, should successfully compete with interference UV and deep UV filters.

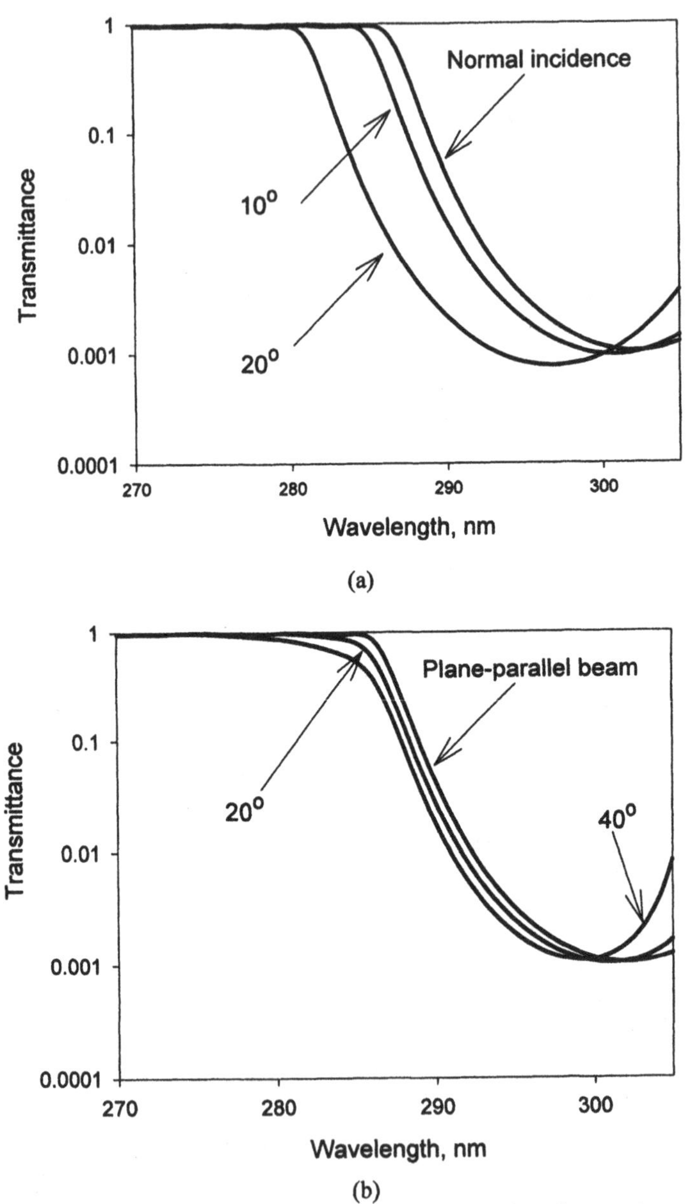

Figure 5.19. Numerically calculated transmission spectra through an interference short-pass filter (a) a through a five-layer coated MPSi layer (b) for different angles of incidence.

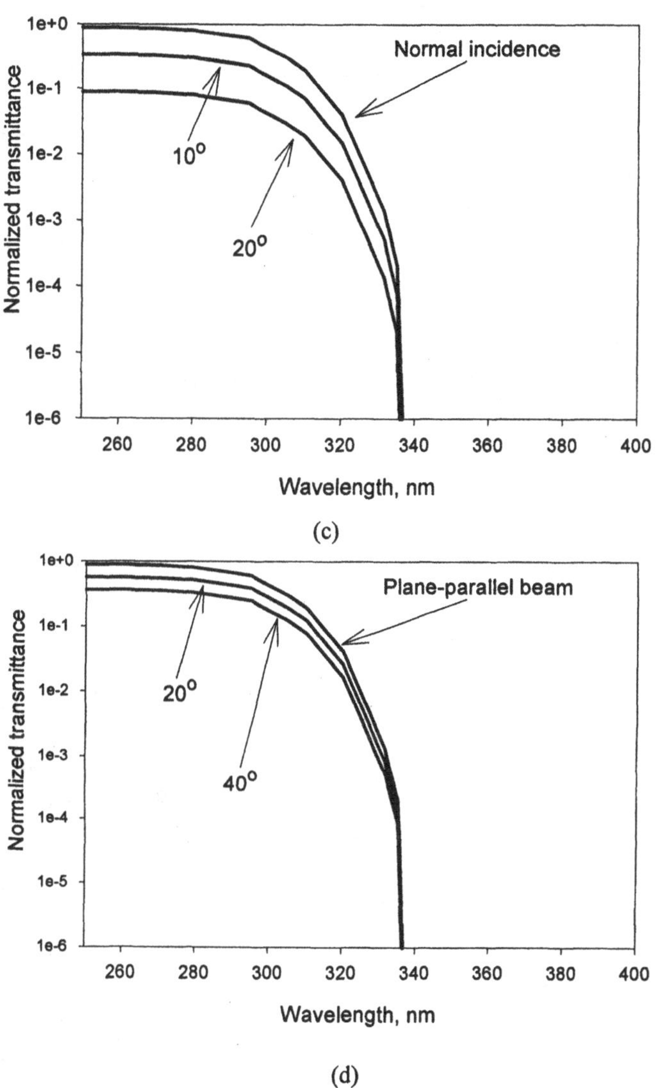

(d)

Figure 5.19 (*continued*). Numerically calculated transmission spectra through an interference short-pass filter (c) and through a five-layer coated MPSi layer (d) for different diversions of Gaussian beam.

Although in the discussion above we have considered an ordered MPSi array; in principle it does not have to be ordered to provide short-pass spectral characteristics. As will be discussed in section 5.5, the ordered MPSi array requires additional manufacturing steps compared to an unordered MPSi layer case. However, an unordered MPSi array has some disadvantages. Particularly, for an unordered MPSi array it is impossible to create uniform size pores. In the nonuniform pores case, the phase relationship between the leaky waveguide modes in individual leaky waveguides can be completely lost, resulting in diffused scattering of the output beam (see figure 5.20). The optical output in this case will fill all the solid angles, determined by the *numerical aperture* (or, in other terminology, the *acceptance angle*) of the leaky waveguides, regardless of the input direction. Although such MPSi layers are totally useless as filters for imaging; they still will be suitable for applications requiring filters if attached directly to the front facet of a photodetector.

In the case of perfectly uniform leaky waveguides, the relative phases at the coupling interface will be transferred to the output facet, resulting in deviation of the output beam to a direction exactly parallel to the input beam. Moreover, the intensity distribution exhibited across the input beam is expected to be reproduced at the output with a spatial resolution comparable to the pore-to-pore distance.

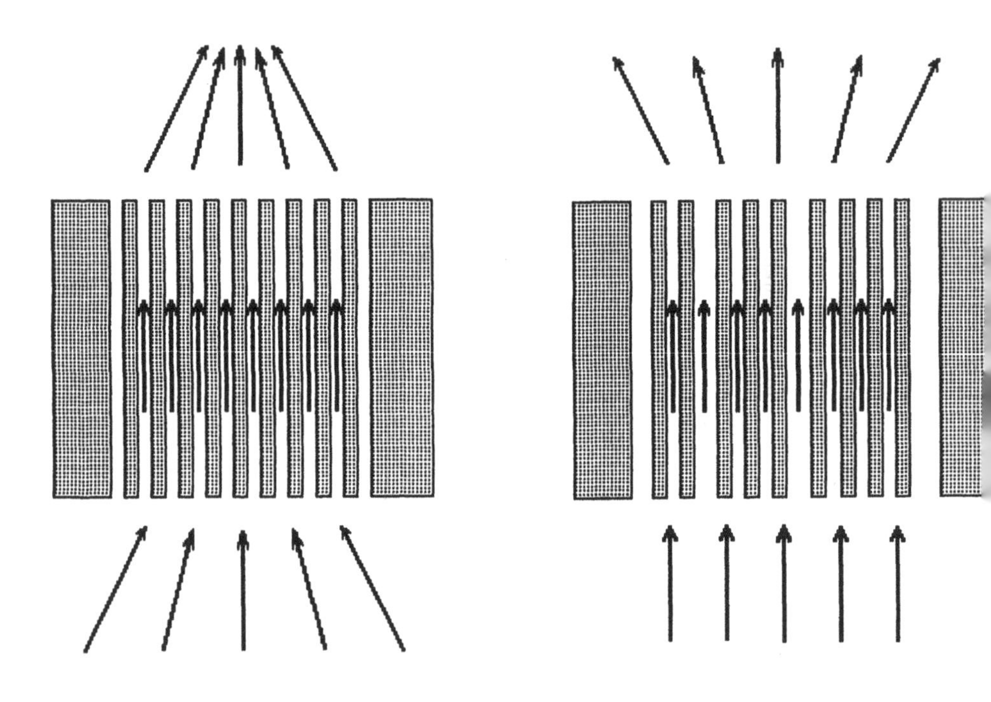

Figure 5.20. Reemitting of light by the second interface of an MPSi filter. Two limiting cases of transmission (with preserved coherence and with totally lost coherence) are shown.

Although there is much more that can be done with MPSi layer design; let us stop now and switch to the manufacturing of MPSi layers, since it is dramatically different from any other method used for optical filters fabrication.

5.4 Macroporous Silicon Formation

Macroporous silicon is usually formed through the silicon anodization process. It generally has an unordered structure. However, some methods have been developed to obtain ordered macroporous silicon arrays. First, let us examine the basic properties of random MPSi formation. The methods of ordered MPSi array formation are discussed later.

5.4.1 Unordered MPSi Layer Formation

Porous silicon (PSi) is a system of interconnected silicon islands forming a matrix with a predetermined (20 to 95%) porosity. The size of the pores (or Si islands) can vary in the range from nanometers to hundreds of micrometers. Usually, depending on pore size, PSi is divided into three groups:

- Microporous silicon with pore sizes below 2 nm,
- Mesoporous silicon with pore sizes in the range 2 to 50 nm, and
- Macroporous silicon with pore sizes above 50 nm (MPSi).

MPSi is usually formed by electrochemical etching of Si in a solution containing hydrofluoric acid (HF). The pore size, shape, and depth (as well as ordering) are well controlled [49]. A schematic drawing of a simple anodization chamber is given in figure 5.21.

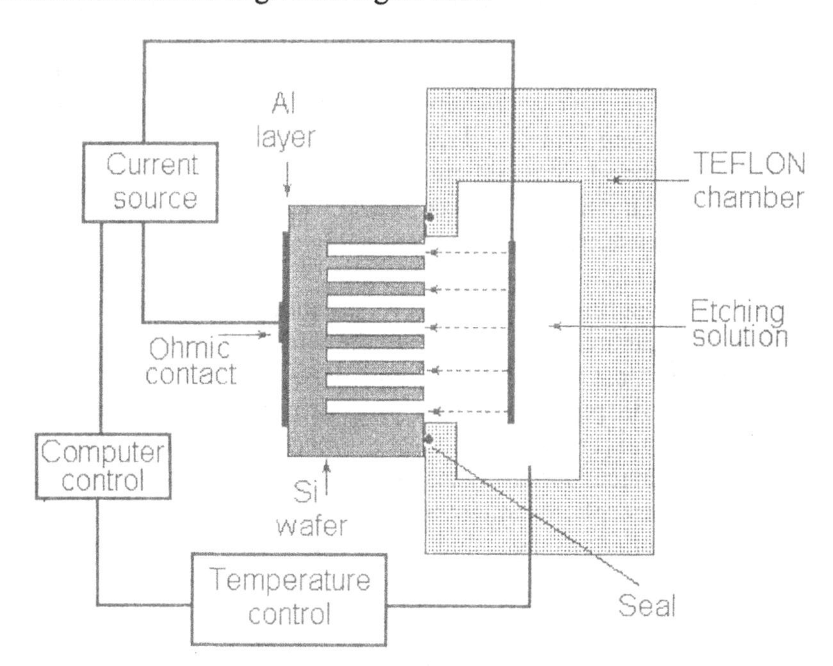

Figure 5.21. A typical silicon anodizing chamber.

The exact dissolution chemistries of silicon are still in question. However, it is generally accepted that holes in the semiconductor valence band are required to supply the current for both electropolishing and pore formation. During pore formation, two hydrogen atoms evolve for every silicon atom dissolved [50]. The hydrogen evolution diminishes as the electropolishing regime is approached (high current densities) and disappears during electropolishing. Current efficiencies are about two electrons per dissolved silicon atom during pore formation and about four electrons in the electropolishing regime [51].

The basic reactions, involved in the anodic etching of silicon in aqueous HF, are

$$Si_{(s)} + 6HF_{(aq)} \rightarrow H_2SiF_{6(aq)} + 2H_2(g) \qquad (5.11)$$

$$Si_{(s)} + O_{2(g)} \rightarrow SiO_{2(s)} \qquad (5.12)$$

$$H_2SiF_{6(aq)} + 8H_2O \rightarrow SiO_{2(s)} + 6H_3O^+_{(aq)} + 6F^-_{(aq)} \qquad (5.13)$$

Reaction (5.11) is the oxidation/reduction reaction between silicon and the hydronium ion, which generates a soluble silicon fluoride complex and evolves hydrogen gas. Equation (5.12) is a side reaction characterized by the oxidation/reduction reaction between silicon and oxygen directly to form silicon dioxide. It is slow, even with a driving force, since both elements have to leave their standard states. The silicon dioxide is soluble in the electrolyte (since it necessarily contains HF) and dissolves in HF, causing a disruption to the etching process. Equation (5.13) is another side reaction: the ion exchange reaction. It again creates silicon dioxide that has the same effect as in (5.12). The silicon hydride bonds passivate the silicon surface unless a hole (positive current carrier) is available.

Although the reactions (5.11) to (5.13) are straightforward; there are many more parameters that complicate the anodic etching of silicon. These reaction parameters can be divided into three groups: silicon wafer parameters, electrolyte parameters, and process parameters. Silicon wafer parameters are crystal orientation, doping type, and density. Electrolyte parameters are HF concentration, types, and concentrations of additives, O^{2-} and O_2 concentrations, pH, conductivity, and viscosity. Process parameters are temperature, applied voltage, and current density. All three groups of reaction parameters have been reported to play significant roles in determining which of the reactions (5.11) to (5.13) are dominant and, as a result, the morphology of the macropores. The results of some of the above reaction parameters are discussed in more detail later in the text.

Theoretically, the silicon anodization process is divided into two stages: (1) nucleation and reorganization processes and (2) stable pore growth. The above process parameters affect each stage differently. Nucleation and reorganization processes have been characterized by the dissolution process during the silicon etching. Stable pore growth has been characterized by a steady-state condition between ionic diffusion in the electrolyte and charge supply from the silicon electrode. These two stages are usually simple to distinguish by monitoring the volt-ampere characteristics of the silicon anodization process.

As an example, figure 5.22 gives the applied voltage as a function of time (constant current mode) for a p-doped silicon wafer with resistivity 66 Ωm cm in contact with electrolyte having the composition {1 [HF] +2 [Alcohol] + 8 [CH_3CN]}. The current density was kept constant at 4.8 mA/cm^2. It is illustrated that the nucleation and reorganization stage usually requires higher voltage and lasts from 5 to 30 minutes depending on particular process parameters. The stable pore growth stage is usually characterized by stable volt-ampere characteristics.

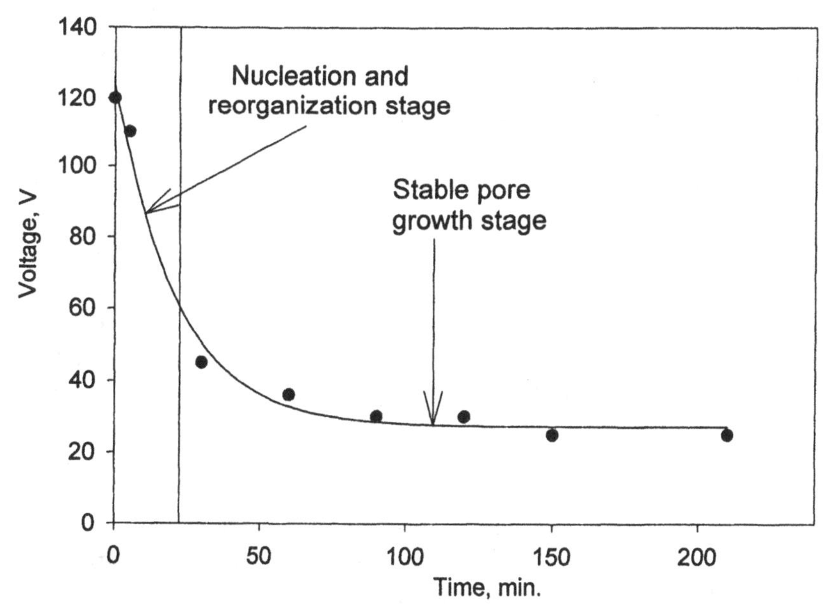

Figure 5.22. Applied voltage as a function of time (constant current mode) for a p-doped silicon wafer with resistivity 66 Ωm cm in contact with electrolyte having the composition {1 [HF] + 2 [Alcohol] + 8 [CH_3CN]} under the current of 4.8 mA/cm^2. The anodized area of the silicon wafer is ~12 cm^2.

Porosity p of the PSi (and MPSi) layer is defined as the volume fraction of voids within the layer. For the MPSi array the porosity p is determined by the etching current density J according to $p = J/J_{PS}$, where J_{PS} is the critical current density at the pore tips [49]. For an orthogonal pattern of pores (cubic symmetry) with a pitch i, the pore size is simply $d = i \times p^{1/2}$ if the cross-section of a pore is approximated by a square. Both doping density and current density are crucial parameters for the pore diameter. The doping density and the pitch control the diameter and morphology of the pores. Branching of the pores occurs for high doping densities, and tapering-off of the pores occurs for low doping densities.

As described above, the macropore diameter depends on the critical current density J_{PS} at the pore tip, which itself depends on the concentration of the bulk electrolyte, the temperature, and the diffusion in the pore [51]. The decrease of electrolyte concentration in the pore is reported to be linear [51] in the approximation of shallow pores.

The maximum obtainable MPSi depth depends on diffusion limitations in the narrow pores and therefore on the electrolyte temperature and concentration. For a given electrolyte composition, concentration and pore geometry, a maximum pore depth exists at which the growth rate becomes minimal at the pore tip and increases at the pore walls. The region of stable pore growth is identified by a constant pore diameter. In general, the lower the HF concentration, the deeper the pore depth obtained; although the particular dependence is different for different electrolyte compositions. However, most electrolytes that are suitable for MPSi formation with a pore cross-section around 1 μm are also suitable for obtaining a desired MPSi layer thickness of 100 to 300 μm. Figure 5.23 presents an experimentally measured MPSi layer thickness as a function of time for an n-doped silicon wafer with resistivity within the range of 0.8 to 1.3 Ωm cm, processed in an electrolyte having the composition {1 [HF] + 2 [Alcohol] + 8 [H_2O]} under the current density of 9 mA/cm^2 and back-side UV illumination of 15 W/cm^2. The macropores were obtained with circular cross-section and diameters around 1 μm. As follows from figure 5.23, not only has the MPSi thickness limit not been reached, but the decrease in the MPSi growth rate, caused by the HF concentration gradient, has not been detected.

MPSi can be formed from both p- and n-type doped Si wafers. A good description of the MPSi formation process on n-type doped (usually phosphorous-doped) wafers can be found in [52]. To get pore diameters of around 1 μm one needs to use an n-type doped silicon wafer with resistivity in the range of 0.5 to 5 Ω cm. Electrolytes, used to produce an MPSi layer on such wafers usually contain a mixture of HF, alcohol, and water. Alcohol (or ethanol) is used to dissolve hydrogen, which is generated during silicon etching (see equation (5.11)). HF concentration is usually defined by the pore size needed and can be anywhere from 2 to 20 weight percents. MPSi formation on n-type doped wafers has some peculiarities. Particularly, backside illumination is required to get good-quality pores. MPSi formation on n-type doped wafers is not simply an electrochemical process but rather a photoelectrochemical process.

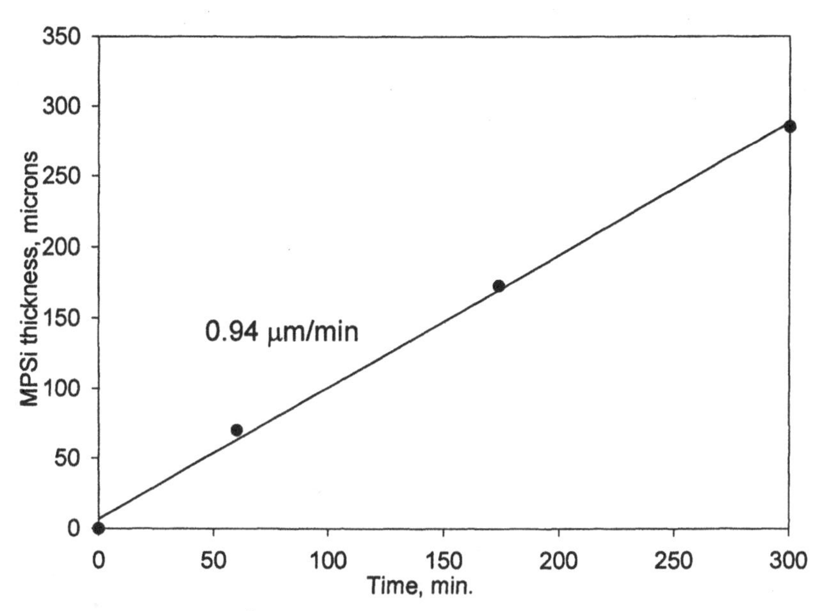

Figure 5.23. MPSi layer thickness as a function of time for n-doped silicon wafer with resistivity within the range of 0.8 to 1.3 Ωm cm, processed in electrolyte having the composition {1 [HF] + 2 [Alcohol] + 8 [H_2O]} under the current density of 9 mA/cm^2 and back-side UV illumination of 15 W/cm^2.

Silicon wafer backside illumination provides an additional supply of holes at the silicon/electrolyte interface. Hence, illumination is needed with wavelengths below the absorption band-edge of silicon. Usually UV lamps are used for this purpose.

Illumination changes the volt-ampere characteristic of silicon/electrolyte contact. In figure 5.24, the current density is given as a function of applied voltage for different values of illumination intensity for an n-doped silicon wafer with resistivity within the range of 0.8 to 1.3 Ωm cm in contact with electrolyte having the composition {1 [HF] + 4 [Alcohol] + 21 [H_2O]}. The area of silicon wafer that has been anodized was ~ 12 cm^2. The voltage needed to get a particular current density generally decreases with the increase of illumination intensity.

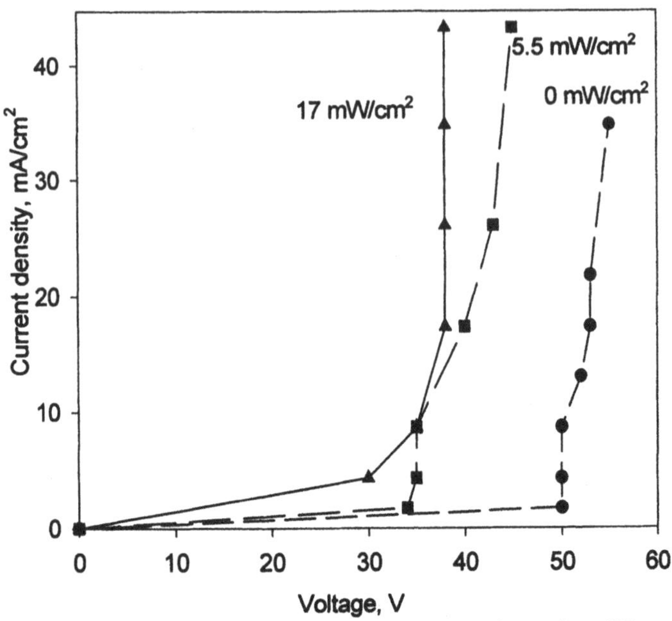

Figure 5.24. Current density as a function of applied voltage for different values of illumination intensity for an n-doped silicon wafer with resistivity within the range of 0.8 to 1.3 Ωm cm, processed in electrolyte having the composition {1 [HF] + 4 [Alcohol] + 21 [H_2O]}. The area of silicon wafer that was anodized is ~ 12 cm^2.

Beyond the illumination needed, n-type silicon has other disadvantages for MPSi fabrication. For a given electrolyte composition and silicon wafer doping type and density, the growth rate is proportional to the local current density and to the local back-side illumination power density. If the current and/or illumination power densities are not constant across the silicon wafer area exposed to the electrolyte, the MPSi layer thickness will be also nonuniform. Processing reasonably large (from 1 to 6 inches round) wafer areas through the anodization step requires a more or less uniform current and/or illumination power densities; the conductive mesh (usually gold, copper, or aluminum) should be created on the back-side of the silicon wafer. This adds another photolithographic and etching step to the overall MPSi fabrication procedures.

P-type doped silicon (usually boron-doped) seem to have significant advantages over n-type doped silicon from the point of view of MPSi manufacturing, since back-side illumination is usually not needed for MPSi growth on p-type doped wafers. However, unlike the n-type silicon wafer case, p-type doped silicon anodization electrolyte usually contains organic addends. According to Christopherson [53], macropore formation on p-type silicon can take place in HF-containing electrolytes with the addition of acetonitrile (MeCN or CH_3CN) [54], diemethyl formamide (DMF or $HCON(CH_3)_2$) [55], diluted HF without organic agents [56], diemethyl sulfoxide (DMSO or $(CH_3)_2SO$) and mixtures of these electrolytes with water.

With organic-addend containing electrolyte the water supplies oxygen and hydrogen during silicon anodization, as outlined in [57]. Diethyleneglycol (DG, or $HOCH_2CH_2OCH_2CH_2OH$) supplies only hydrogen. Also, according to Christopherson [53], the supply of oxygen to the electrolyte/silicon wafer interface should be somewhat restricted, and the supply of reactive hydrogen should be large. The conductivity of the electrolyte, the viscosity and the stability of its components must be taken into account.

Since the pores fabricated in the MPSi layer should be perpendicular to the silicon wafer surface, silicon wafer has to be of <100> orientation. Moreover, the direction of the current flowing through the wafer also has to be perpendicular to the wafer surface, since even for <100> wafer orientation, macropores tend to line up toward current lines. The later condition usually satisfies automatically; however, tilted pores have been observed at the areas of bad electrical contact at the back surface of the silicon wafer (when, for example, it contains silicon dioxide, silicon nitride, or other isolating surface features).

The best transmission through the MPSi layer within its pass band requires the straight pores with constant diameters. The "bad" pores (see figure 5.25) in which diameters change with depth or branching of the pores occurs, were specifically observed in etching experiments with relatively large amounts of available oxygen – usually supplied by the water in the system. Thus, the proportion of water in the electrolyte should be kept low. Extremely straight macropores have been obtained in systems with reduced oxygen and increased hydrogen availability for the interface reactions [53].

The temperature of the electrolyte during silicon anodization also plays an important role in pore wall smoothness. It was found that anodization experiments at temperatures ~2 to 5°C produce a good-quality MPSi layer.

Different types of organic addends in combination with different levels of resistivity of p-type doped silicon wafers produce different ranges of pore sizes and pore-to-pore spacing.

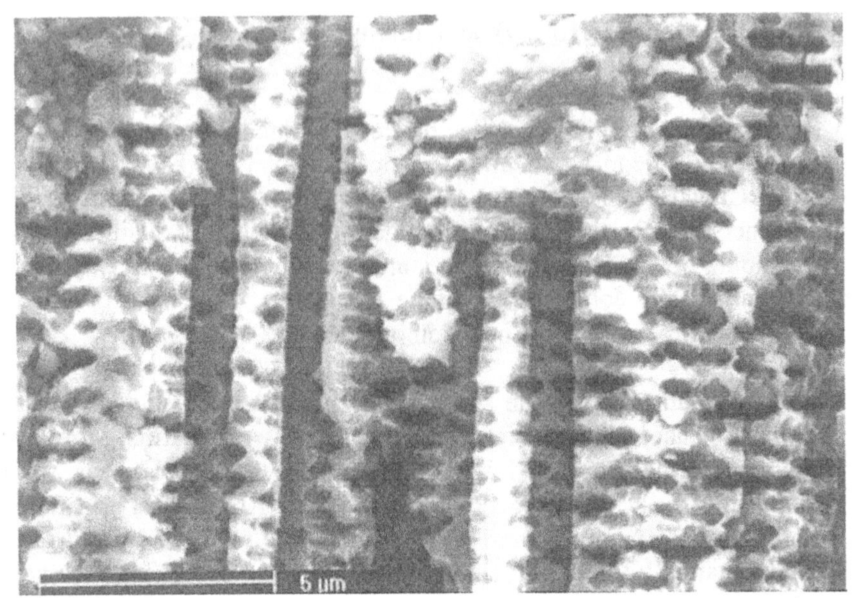

Figure 5.25. An SEM image of the MPSi layer produced under excessive amount of oxygen (through concentration of water) in electrolyte ("bad" pores).

The smallest pore diameters of good-quality MPSi layers (i.e., with smooth, straight macropores) were obtained in DMF containing electrolytes with silicon wafer resistivity in the 5 to 15 Ωcm range.

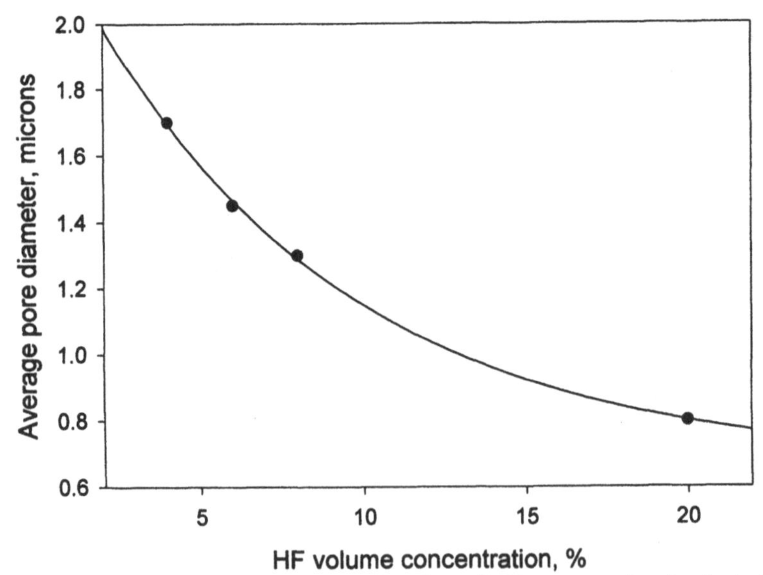

Figure 5.26. Average pore diameter as a function of HF concentration in electrolyte for p-doped silicon wafer with resistivity within the range of 5 to 20 Ωm cm, processed in electrolyte having the composition {x [HF]+(1-x) (1 [Alcohol]+4 [$HCON(CH_3)_2$])} under the current density of 4.8 mA/cm^2.

Figure 5.26 gives an average pore diameter as a function of HF concentration in electrolyte for a p-doped silicon with resistivity within the range of 5 to 20 Ωm cm, processed in electrolyte having the composition {*x* [HF]+(1-*x*) (1 [Alcohol]+4 [$HCON(CH_3)_2$])} under the current density of 4.8 mA/cm^2. For the DMF-based electrolyte and silicon wafer resistivity range, given above, good-quality pores were obtained for HF volume concentrations below ~ 10%. For the electrolytes containing a higher concentration of HF, the MPSi layer quality was not sufficient for optical filtering due to the degradation of pore walls quality. For lower than three volume percent HF concentration, the etch rate was found to be too low (less than 0.3 microns per minute). Lower resistivity (less than 5 Ωcm) silicon wafers with low concentrations of HF in electrolyte were also found to produce a poor quality MPSi layer.

DMF-contained electrolytes were found to have some advantages and disadvantages as a fabrication base for MPSi-based optical filters. The advantages of DMF-based electrolytes include absence of nanopore coverage of macropore walls, relatively small pore diameter and pore-to-pore distance obtainable using such an electrolyte. However, it was found that considerably high stresses were developed between the MPSi layer and the unetched silicon wafer part. Particularly, it was found that for anodization of silicon openings with areas more than 1 in^2, the MPSi layer had a tendency to break-off the silicon wafer when its thickness exceeded 120 to 150 μm, while the thickness for 2 in^2 openings was limited by 80 to 00 μm.

Acetonitrile showed more promise than DMF as an electrolyte base for getting a better-quality MPSi layer. It was found that when used with an MeCN, silicon resistivity should exceed approximately 15 Ωcm. For lower resistivity silicon, the MPSi layer that was formed was found not to be sufficient quality (the pore wall smoothness was poor). For silicon wafer resistivity higher than 15 Ωcm, no MPSi layer break-off (like in the DMF case) were observed, and MPSi layers with up to 300 μm thickness were obtained over 3 in^2 wafer openings. Figure 5.27 give MPSi layer thickness as a function of time for a p-doped silicon wafer with resistivity within the range of 40 to 80 Ω cm processed in electrolyte having the composition {1 [HF] + 2 [Alcohol] + 8 [CH_3CN]} under the current density of 4.8 mA/cm^2.

The best-quality MPSi layers with reasonably small pore diameters were obtained with MeCN-based solutions for p-type doped silicon wafers having resistivity in the range of ~30 to 80 Ωcm. The higher-resistivity wafers (up to 200 Ωcm) also produced good-quality MPSi layers; although pore diameters in this case were too large for deep UV filter design.

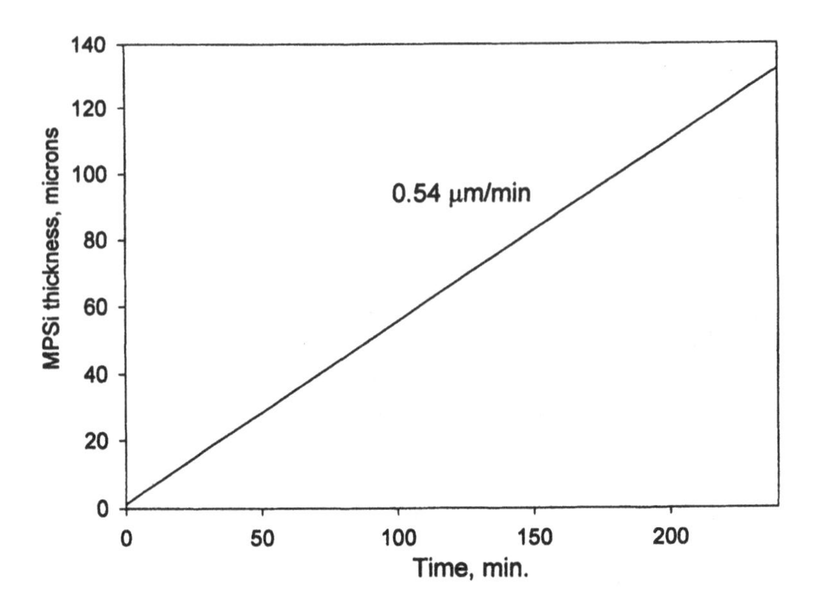

Figure 5.27. MPSi layer thickness as a function of time for a p-doped silicon wafer with resistivity within the range of 40 to 80 Ω cm processed in electrolyte having the composition {1 [HF]+2 [Alcohol]+8 [CH_3CN]} under the current density of 4.8 mA/cm^2.

Figure 5.28a presents the average pore diameter (in random pore array) as a function of silicon wafer resistivity for a p-doped silicon wafer processed in electrolyte having the composition {3 [HF]+5 [Alcohol]+16 [CH_3CN])} under the current density of 4.8 mA/cm^2, while figure 5.28b presents the average pore-to-pore distance as a function of silicon wafer resistivity for the same conditions.

As for the DMF case, HF concentration in MeCN-based electrolyte is playing an important role in average pore diameters and average pore-to-pore distance as silicon resistivity. Figure 5.29 illustrates average pore diameter (a) and average pore-to-pore distance (b) as a function of HF concentration in electrolyte for a p-doped silicon wafer with resistivity within the range of 40 to 80 Ωm cm processed in electrolyte having composition {x [HF]+(1-x) (1 [Alcohol]+4 [CH_3CN])} under the current density of 4.8 mA/cm^2.

It should be noted that, for high HF concentrations (10% and above), the MPSi layer quality was quite poor. It follows from figures 5.28 and 5.29 that the lowest average pore diameter for a good-quality MPSi layer obtainable with an MeCN-based solution is around 0.9 μm while the minimal average pore-to-pore distance is around 1.7 μm.

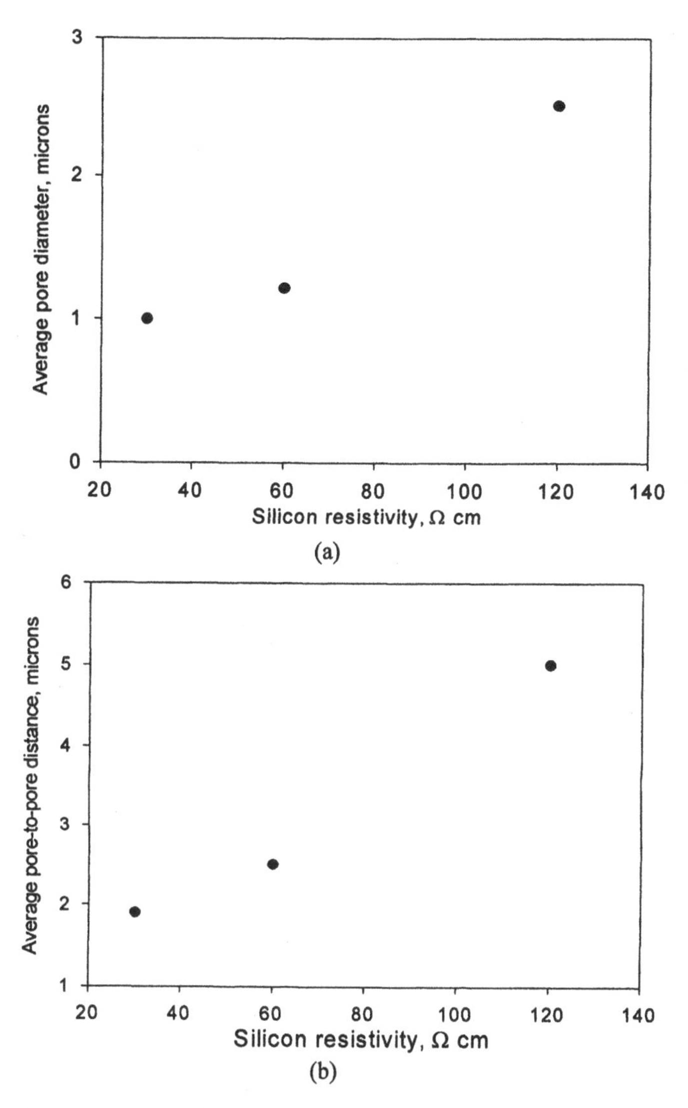

Figure 5.28. Average pore diameter (a) and average pore-to-pore distance (b) as functions of silicon resistivity for p-doped silicon wafer processed in electrolyte having the composition {3 [HF]+5 [Alcohol]+16 [CH_3CN])} under the current density of 4.8 mA/cm^2.

Despite the previously mentioned advantages for fabricating good-quality MPSi layers, MeCN-based electrolytes also have some disadvantages. The main disadvantage is that the micropore coverage of macropores always takes place with MeCN-based electrolytes. Micro-roughness of the pore walls can be a significant disadvantage for MPSi layers that are used as a base for deep UV filters. It should be noted that micropore coverage can be removed from the macropore walls by thermal oxidation of the MPSi layer followed by HF etching-off of the grown silicon dioxide. However, it adds complexity to the MPSi fabrication process.

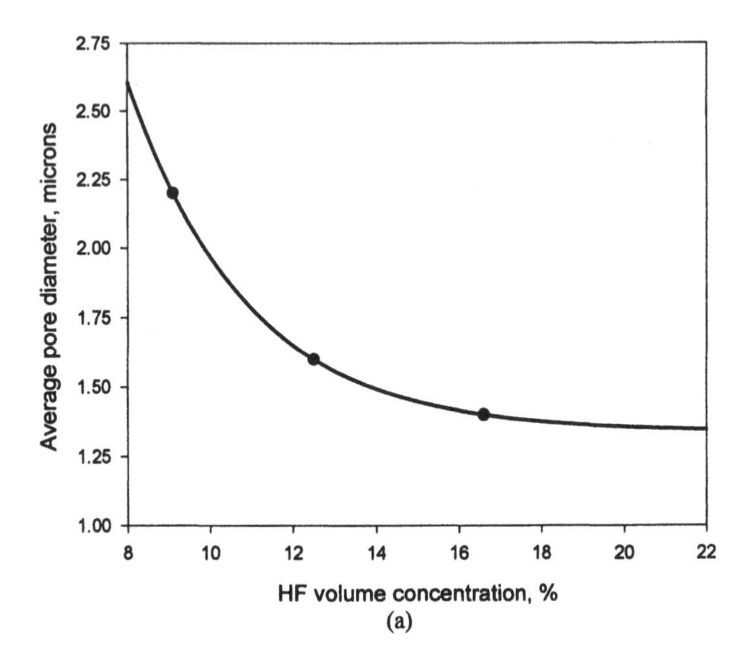

(a)

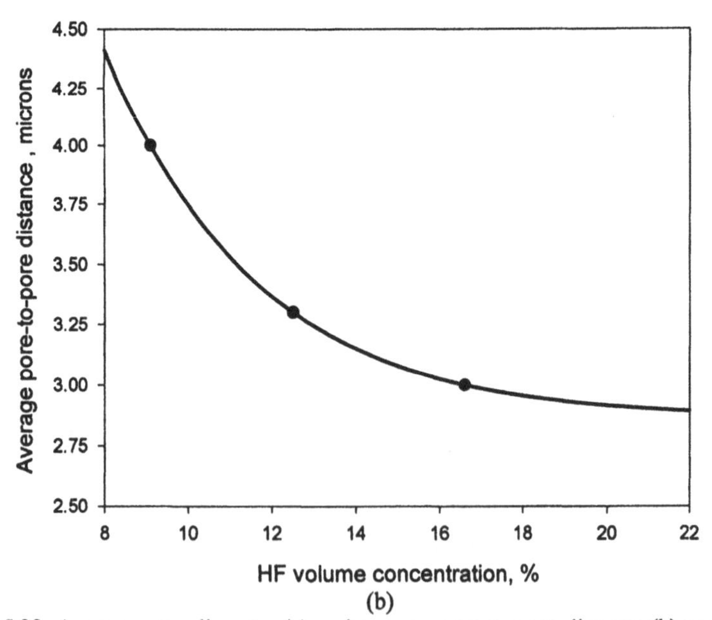

(b)

Figure 5.29. Average pore diameter (a) and average pore-to-pore distance (b) as functions of HF concentration in electrolyte for a p-doped silicon wafer with resistivity within the range of 40 to 80 Ωm cm processed in electrolyte having the composition {x [HF]+(1-x) (1 [Alcohol]+4 [CH_3CN])} under the current density of 4.8 mA/cm^2.

It was found that diemethyl sulfoxide (DMSO or $(CH_3)_2SO$) as a base of electrolytes offers the absence of micropore coverage of macropore walls (the same as DMF) together with a low level of stresses between the MPSi layer and the unetched part of the silicon wafer (the same as MeCN). For p-doped silicon wafers with resistivity in the range of 40 to 80 Ωcm, good-quality MPSi layers have been obtained with more than 200 μm thicknesses when anodization was done in DMSO-based electrolytes. No break-off of the MPSi layer was observed, which indicates that the above-mentioned stresses are within the acceptable range. An SEM image of a macropore wall cross-section is given in figure 5.30. The resistivity of the anodized wafer was 57 Ωcm, the electrolyte composition was 1 [HF]+2 [Alcohol]+10 [$(CH_3)_2SO$], and the current density was 5 mA/cm^2. Although the pore wall is not perfectly smooth; the roughness in the range of 5 to 10 nm is within the acceptable range for UV filters (see the discussion in the previous section). The best-quality MPSi layers with reasonably small pore diameters were obtained with DMSO-based solutions for p-type doped silicon having resistivity in the range of ~30 to 80 Ωcm.

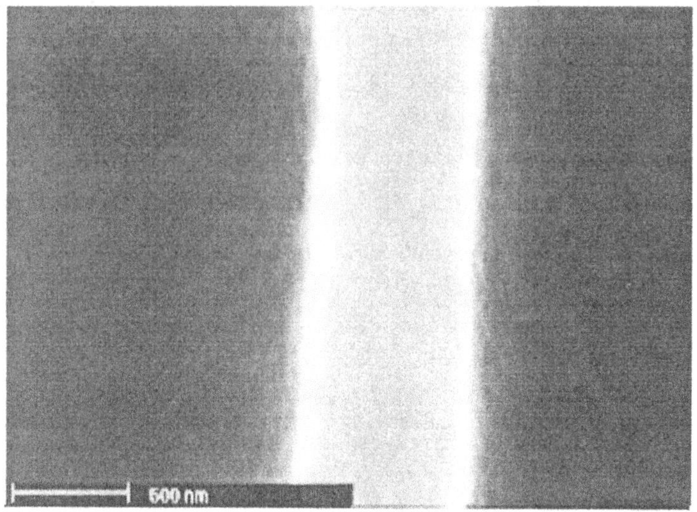

Figure 5.30. A high-resolution SEM image of macropore wall cross-section.

As with the MeCN case, higher-resistivity wafers (up to 200 Ωcm) also produced good quality MPSi layers. However, pore diameters were too large for deep UV filter design. Optimum electrolyte composition was found to be {x [HF]+(1-x) (1 [Alcohol]+4 [$(CH_3)_2SO$])} with x in the range of 2 to 10 volume percents. The optimum current density was found to be in the range of 3 to 10 mA/cm^2.

5.4.2 Ordered Macroporous Array Formation

As mentioned above, the MPSi layer formed on an initially smooth silicon surface does not provide any long-range order. The macropores in such layers are usually disposed quite randomly with respect to each other. Such MPSi arrays can be characterized by average pore-to-pore distance $\bar{l}_{pp} = \frac{1}{n}\sum_{i=1}^{n} l_i$ (where l_i is the ith measured pore-to-pore distance and n is the number of measured pore-to-pore distances) and the dispersion of the pore-to-pore distance $\sigma_{pp} = \frac{1}{n}\sqrt{\sum_{i=1}^{n}\left(\bar{l}_{pp} - l_i\right)^2}$. Figure 5.31 presents an SEM image of a random MPSi array obtained by anodization of p-type doped silicon with resistivity of 12 Ωcm in an electrolyte, having the composition {1 [HF] + 2 [Alcohol] + 10 [DMF]} processed at the current density of 4.8 mA/cm^2. As was mentioned in the previous section, average pore-to-pore distance can be controlled through the electrolyte composition, wafer resistivity, applied voltage, and current during silicon anodization reaction.

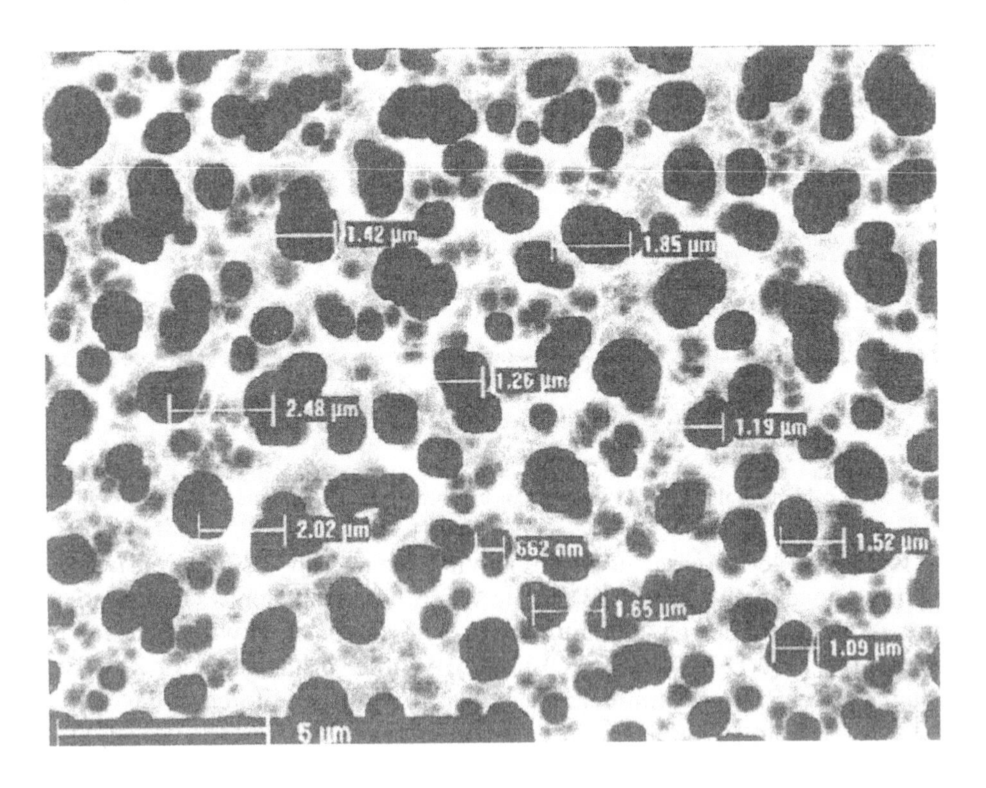

Figure 5.31. An SEM image of a random MPSi array.

The dispersion of pore-to-pore distance can also be controlled to some extent. For example, it can be kept at reasonably low levels (in the range of 5 to 15% of average pore-to-pore distance) by performing silicon anodization at temperatures slightly exceeding the freezing temperature of electrolyte. By further optimizing the process conditions (like current density and applied voltage), it is possible to obtain fairly closely packed macropores providing quite reasonable overall transmission through the MPSi layer at its pass-band. However, as discussed above, the coherence of the light transmitted through a random MPSi array will be completely lost, leading to the scattering of the light at the second interface of the MPSi layer. While it is not a problem for some users, it is completely prohibitive for other applications.

Other important parameters of random MPSi arrays are average macropore diameter $\bar{d} = \frac{1}{n}\sum_{i=1}^{n} d_i$ (where d_i is the ith measured macropore diameter and n is the number of measured macropore diameters) and the dispersion of the macropore diameter $\sigma_d = \frac{1}{n}\sqrt{\sum_{i=1}^{n}\left(\bar{d} - d_i\right)^2}$. Just as for pore-to-pore distance, it is possible to keep the dispersion of macropore diameters reasonably low (in the range of 5 to 15% of average macropore diameter) by optimizing anodization process conditions. However, the transmission spectrum of MPSi layer, especially with pore walls coated by the multilayer, is very sensitive to pore diameter. Unfortunately, the best-obtained dispersions of the pore diameters are not good enough to get the optimized performance of the MPSi layer as a filter. Better control of the pore diameters is possible in ordered MPSi arrays only.

For ordered macropore patterns, initial pits are required as starting points. The pits make it possible to control the location of the macropores. The initial pits require a sharp tip in order to collect electronic holes, which work as a trigger for the chemical reaction; although sharpness is not a very strict requirement. The pits can be formed by either isotropic (or anisotropic) wet etching or a reactive ion etching (RIE), followed by electrochemical etching. Lehmann [17], [52] suggested alkaline etching. Ohji [58] showed that HF, HNO_3, and acetic acid are also suitable. In the author's experience, KOH or TMAH solutions are better, since electrochemical etching always starts at the center of each initial pit due to the inverse pyramid shape of the pit, which is an efficient generator of electronic holes.

Etch-pit formation is a multistage process. Usually the silicon wafer surface to be anodized is coated by a thin (40 to 500 nm) layer of chemically resistant material. Silicon dioxide (either thermally grown or sputtered) or silicon nitride (deposited by CVD) are common materials for these purposes. Next, the photoresist mask is formed on top of the chemically resistant layer with openings properly sized and spaced through standard photolithography process.

The chemically resistant layer is etched through the photoresist mask until the silicon surface is reached. For large-period MPSi arrays (more than ~ 5 μm pore diameters), chemical etching (hydrofluoric acid for the silicon dioxide layer or hot phosphoric acid for the silicon nitride layer) can be used at this stage. For smaller features, however, reactive ion etching (RIE) or ion milling (IM) is required.

An SEM image of an etch-pit mask in a silicon dioxide layer covering the front surface of a silicon wafer is given in figure 5.32. In the next step, the wafer with the etch pit mask in a chemically resistant layer is placed into a hot KOH or TMAH (tetramethyl ammonium hydroxide) solution. The KOH concentration should be around 40%, and the temperature should be in the 70 to 80°C range to get the best-quality etch-pits. Due to the anisotropic nature of silicon etching in a KOH solution (the etch rate in the (100) direction is by more than order in magnitude faster than it is in the (111) direction), the etch pits will have an inverse pyramid shape with sharp ends. It is critical to have the etch-pit mask in the chemically resistant layer etched all the way through to the silicon, since otherwise (if a thin layer of chemically resistant material is left in the openings), instead of a single inverse pyramid, one can get several randomly placed pyramids inside each etch-pit opening.

After etch pits are etched into the silicon surface, the chemically resistant layer should be removed from the silicon wafer by chemical etching. At this stage the wafer is ready for anodization. It is crucially important to match the average pore-to-pore distance typical for an anodization process parameters to the etch pit array. Mismatch can cause a complete loss of order in the MPSi array. Figure 5.33 gives an SEM image of an MPSi array for the case of a strong mismatch between etch pit array period and average pore-to-pore distance. The process parameters were as follows: the silicon was of p-type and resistivity around 10 Ωcm, and the electrolyte had the composition {1 [HF] + 2 [Alcohol] + 10 [DMF]}, processed at the current density of 4.8 mA/cm^2. The etch pit period in this picture was 5 μm, while the preferred pore-to-pore distance (which is equal to the average pore-to-pore distance for an unpatterned silicon surface) for such a combination of silicon wafer and electrolyte was about 1.8 μm (as discussed in the previous section of this book). Even though the etch pits were of good shape, such a strong mismatch between the preferred pore-to-pore distance and the etch pit array period caused multiple-pore nucleation within each etch pit.

Figure 5.32. (a) An SEM image of an etch-pit mask in an SiO_2 layer covering the front surface of a silicon wafer; (b) etch-pits etched into a silicon surface.

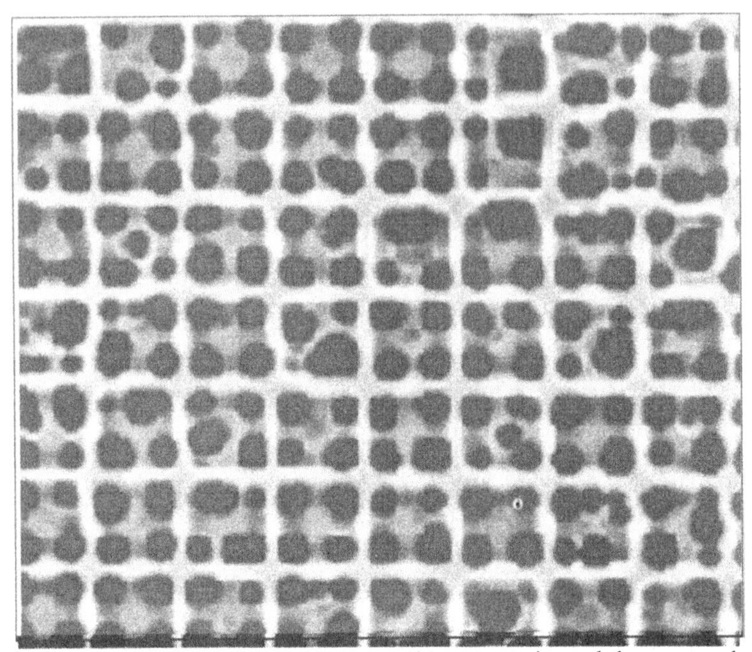

Figure 5.33. An SEM image of an MPSi array with strong mismatch between etch pit array period and preferred pore-to-pore distance.

Figure 5.33 illustrates an extreme case of the mismatch. For lower-level mismatch (the difference between preferential pore-to-pore distance and the etch pit array period between 10 and 50%) the situation is quite different from that of figure 5.33. In this case the majority of macropores nucleates exactly at the etch pit tips. However, the initial order of macropores is usually not preserved through the MPSi thickness. The maximum obtainable thickness of an MPSi layer with the preserved order depends on the level of the mismatch. For lower than an approximately 10% mismatch between the preferential pore-to-pore distance and the etch pit array period, the maximum obtainable thickness of an ordered MPSi array exceeds 200 μm, which is sufficient for MPSi base spectral filter fabrication (see the discussion in the previous section of this chapter).

Adjusting silicon anodization parameters usually requires several probe runs on unpatterned silicon wafers. However, after the optimization stage is completed, the anodization parameters can be used with a particular etch pit mask as long as needed (i.e. as long as the silicon wafer resistivity is kept constant, excellent repetition of the etch pit pattern to pore pattern was observed). Figure 5.35 presents an SEM image of ordered MPSi array. In figure 5.35a the side view of the MPSi array is given, while in figure 5.35b the top view of the same array is presented. The MPSi array was obtained through anodization of a p-doped silicon with resistivity in the range of 10 to 13 Ωcm in the electrolyte having the composition {1 [HF] + 2 [Alcohol] + 10 [DMF]} processed at the current density of 4.8 mA/cm^2. The good quality of etch pit geometry transfer is clearly illustrated.

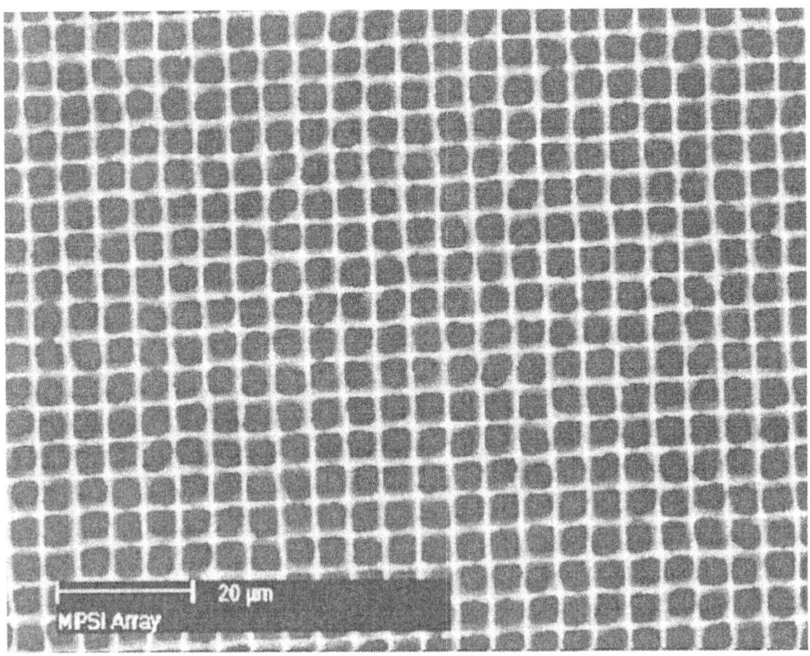

Figure 5.34. An SEM image of an MPSi array with an etch pit array period being about preferred pore-to-pore distance.

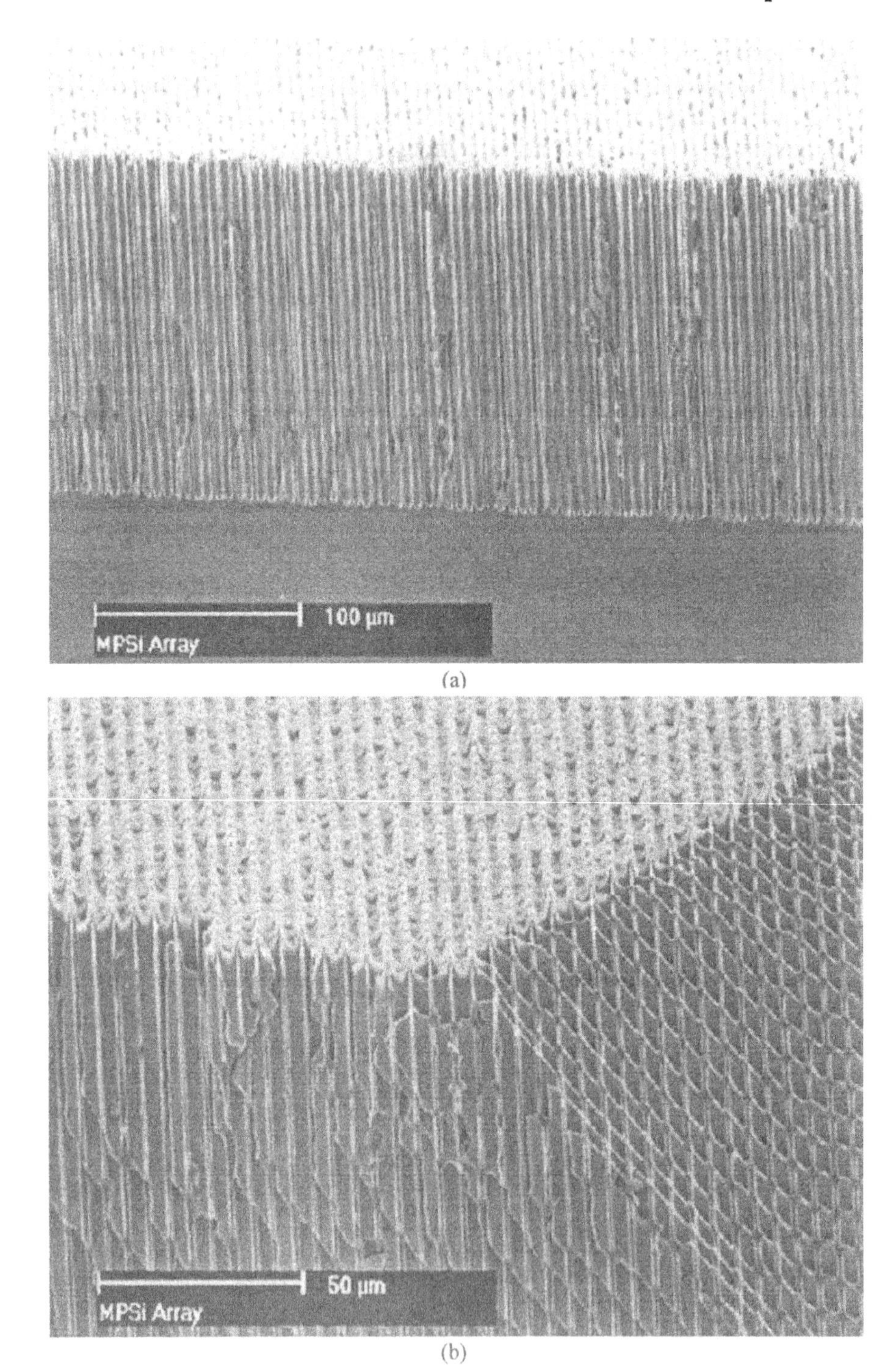

Figure 5.35. SEM images of an ordered MPSi array: (a) side view; (b) tilted view.

5.4.3 Freestanding Macroporous Array Formation

The electrochemical process forms dead-end pores, as illustrated in figure 5.36. As discussed in previous sections of this book, an MPSi layer has to be freestanding in order to serve as a short-pass filter for deep UV wavelengths, since silicon is opaque below ~1000 nm. In order to obtain pores etched through the wafer, the remaining nonporous substrate at the backside of the wafer should be removed. This can be done through chemical etching [17] or through deep reactive ion etching (DRIE).

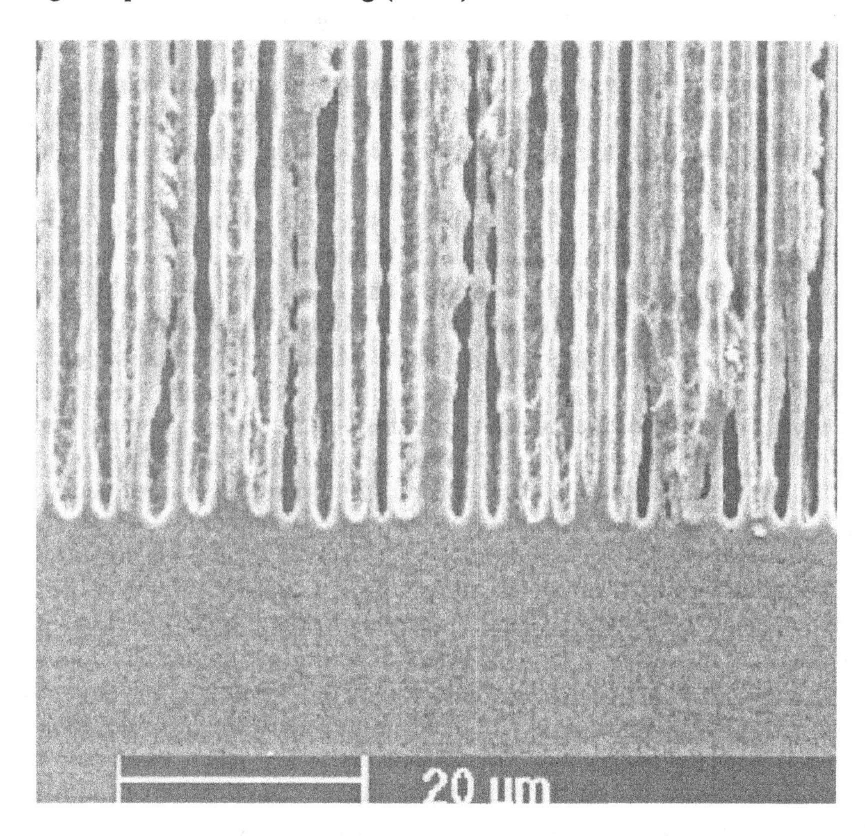

Figure 5.36. An SEM image of macropore ends.

Chemical etching can be performed in hot KOH or TMAH solutions (as in the previous section). However, there are several issues that should be taken into account in this case. The first and most important is that macropore walls are made of the same material as the unetched part of the wafer – from silicon. Since the pore walls are not more than 1 micron thick while the rest of the wafer to be removed is 200 to 300 μm thick, the MPSi layer will be dissolved in the etching solution several hundreds times faster than the rest of the wafer. In [17], it was shown that coating the pore walls with a thin (~100 nm) layer of silicon nitride, which is known to be resistant to KOH or TMAH solutions, can solve this problem. ***Low-pressure chemical vapor deposition*** (***LPCVD***) can be used to deposit the silicon nitride layer onto the macropore walls. The silicon wafer should be coated on both sides with a silicon nitride layer. The filter opening should be defined in the silicon nitride layer from the back side of the wafer. It can be done with a photolithography step followed by RIE of silicon nitride through the photoresist opening. Unfortunately, silicon nitride growth in an LPCVD process is much more involved than thermal growth of silicon oxide.

A silicon dioxide layer (particularly a thermally grown silicon dioxide layer) can be used instead of a silicon nitride layer for protection of the MPSi layer during the chemical removal of the rest of the wafer. However, the selectivity of TMAH and KOH are considerably worse for silicon dioxide than for silicon nitride. In fact, the ratio between the silicon etch rate and silicon dioxide etch rate rarely exceeds 100 (for thermally grown oxide). It means that 100 to 200 nm of the silicon dioxide layer on the pore walls will not protect the MPSi layer from degradation while 200 to 300 μm silicon is removed from the back side of the wafer. The solution to this problem is to protect the front surface of the silicon wafer from contact with the silicon etching solution. In such an arrangement the MPSi layer will be exposed to the KOH or TMAH solution for only a short period of time. Freestanding MPSi layers of a good quality were obtained using such a method. Figure 5.37 gives an SEM image of a freestanding cleaved MPSi layer. This method is clearly the simplest and the cheapest one from the standpoint of necessary equipment. However, there are several other issues that should be carefully addressed. First, it is known that thermally grown silicon dioxide creates a lot of stress both within itself and in the adjacent area of silicon wafer. It sometimes leads to the bending and weakening of the mechanical robustness of the freestanding MPSi layer. The stresses are highest around the edges of the freestanding area of such a wafer. It set the limits on the maximum obtainable area of freestanding MPSi film. The size of this area depends on the thickness of the MPSi layer (the thicker the MPSi layer, the larger area that could be obtained), on porosity *p* (the higher the porosity of the MPSi layer, the smaller the area that could be obtained), and on thermal oxidation conditions. The maximum obtainable area of a good-quality freestanding MPSi layer can be anywhere from 1 in^2 for thin high-porosity MPSi films (50 μm or less) to more than 2 in^2 for 200 μm 50% porosity MPSi layer.

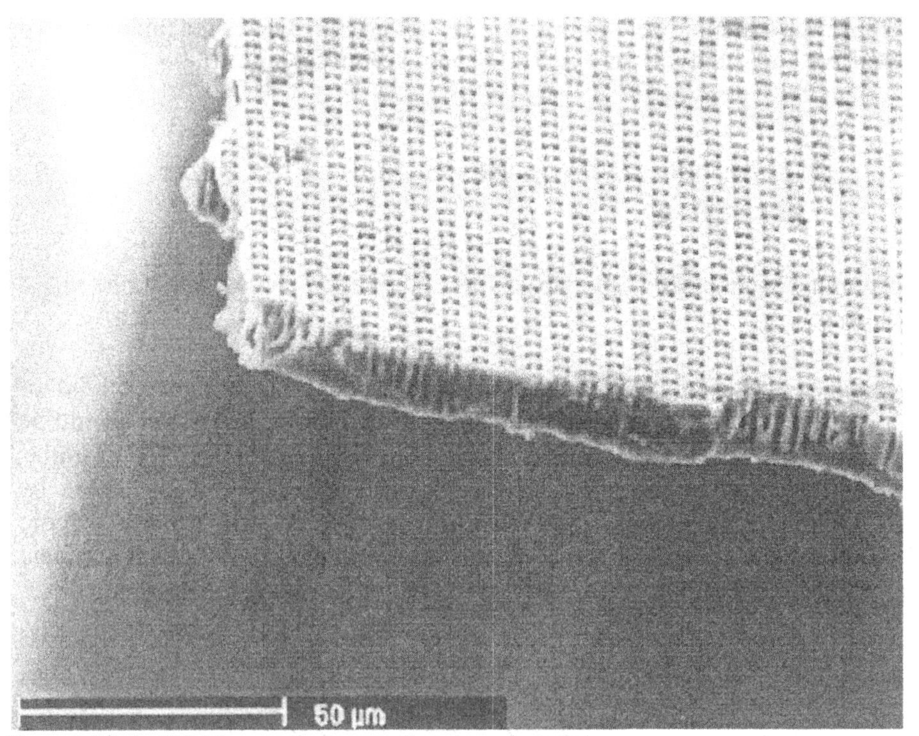

Figure 5.37. An SEM image of a freestanding cleaved MPSi layer.

Another issue that should be addressed when using such a method of freestanding MPSi formation is the pressure difference between the back side of the silicon wafer, which is exposed to etching solution, and the unexposed front side of the wafer. If pressure drop is not compensated, it leads to cracking of the MPSi layer during backside etching, especially for large-area low-thickness MPSi layers.

One issue that is common to all methods of chemical removal of the unanodized part of silicon wafer is the etching-caused roughness of the back surface of the wafer. To suppress it, the initial silicon wafer should be double-side polished. However, even in this case, etching-caused roughness of the back surface of the MPSi layer is usually observed. To suppress it, methanol or other addends are inserted into the KOH or TMAH solution. Still, after the removal of 200 to 300 μm of the silicon layer even in an optimized solution, the back surface of the MPSi layer usually has roughness on the order of several hundred nanometers. This roughness can be suppressed by mechanical polishing of the back side of the MPSi layer. It can be done only after filling the pores with some kind of reinforcing solvable material, for example, wax.

The most straightforward method of back-side silicon removal is *deep reactive ion etching* (*DRIE*). DRIE is 10 to 20 times faster than chemical etching and can be well-controlled. However, the DRIE machine is very expensive and relatively rare in R&D and research facilities.

5.5 Macroporous Silicon-Based Short-Pass Filter Fabrication

To summarize, the MPSi short-pass filter fabrication process should be as follows:

A single-crystal (100) orientation silicon wafer 11 (see figure 5.38a) of doping type and density chosen according to the above discussion should be covered with a thin film of chemically resistant material 11, for example, thermally grown or sputtered silicon dioxide, or silicon nitride grown by LPCVD. Next, the etch-pit mask should be produced in the layer 12 (figure 5.38b) with the assistance of a conventional photolithography and subsequent chemical of RIE etching of layer 12 through the photoresist mask.

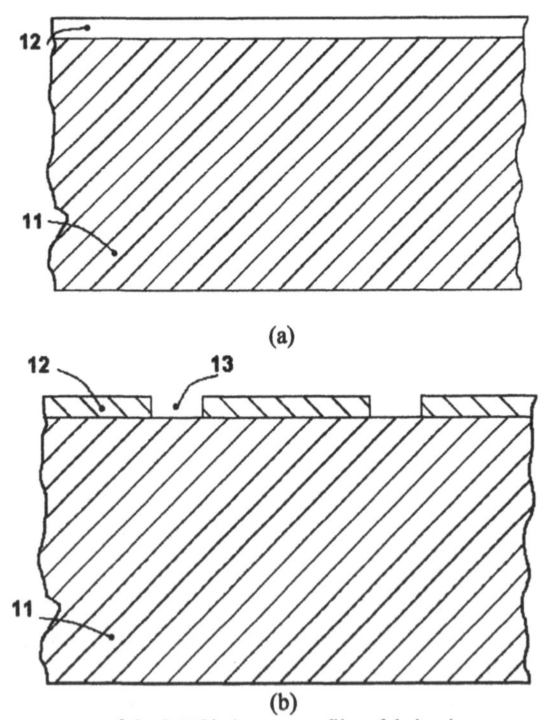

Figure 5.38a-g. A summary of the MPSi short-pass filter fabrication process.

Next, the etch-pit mask should be transferred from the layer 12 into silicon wafer surface 11 (figure 5.38c) by chemical etching with subsequent removal of the chemically resistant layer 12. Next, the first surface (with etch-pits) of the substrate 11 should be brought into contact with an electrolyte (figure 5.38d), the composition of which is carefully chosen according to silicon wafer characteristics and etch-pit array parameters (see previous discussion). The substrate wafer 11 should be connected as an anode. The current density and illumination (if needed) should be applied between the substrate wafer 11 and the cathode placed in the electrolyte. Through the anodization process, the macropores 15 will be grown starting from the tips of the etch pits. The silicon anodization should be performed during the time required to form a macroporous layer with thickness predetermined by filter design considerations. This time can be estimated as a ratio *t/GR*, where t is the desired MPSi layer thickness and GR is the MPSi growth rate, which can be found through calibration runs. After the anodization process is complete, the silicon wafer is removed from the electrolyte. The wafer should be carefully cleaned to ensure that the electrolyte is removed from the deep macropores. If organic addends were used in the electrolyte, Piranha acid cleaning can be required.

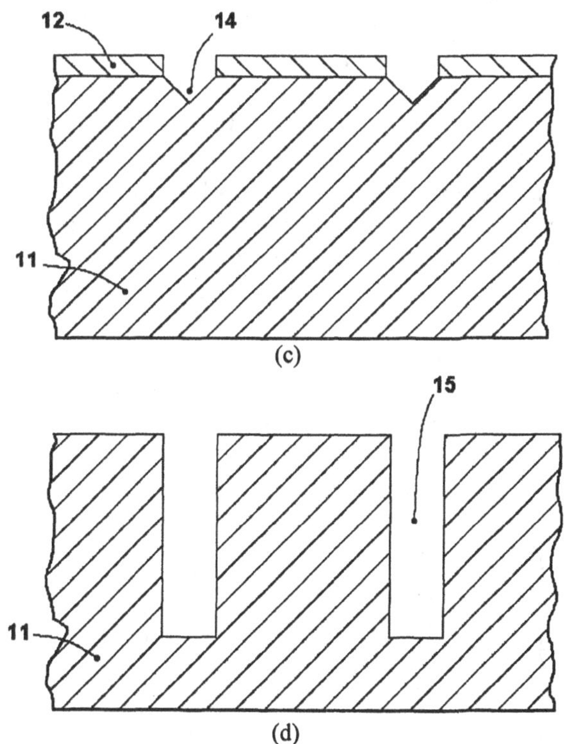

(c)

(d)

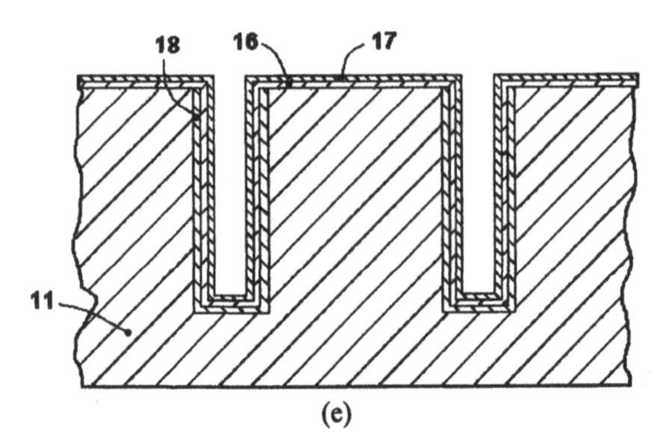

(e)

Next, the first surface of the silicon wafer 11 is covered uniformly by the one or more layers of transparent material(s) (figure 5.38e). The first layer 18 can be grown by the thermal oxidation of silicon wafer 11 in the oxygen atmosphere at temperatures of 950 to1300°C. The thickness of such a silicon dioxide layer can be well controlled by the time of oxidation since this thickness at each temperature is proportional to the oxidation time. In order to reduce oxidation-caused stress in the MPSi layer, the wafer can be annealed in nitrogen atmosphere at temperatures of 400 to 800°C. Although thermal oxidation is a cheaper technique, the CVD and especially LPCVD would produce a low-stress layer, which can be important. All other layers (16, 17 of figure 5.38e) should be deposited by CVD or LPCVD. These layers (16, 17, 18) can be of silicon dioxide, silicon nitride, or any other material sufficiently transparent in UV and deep UV ranges.

Next, the portion of the silicon wafer 11 that does not have the MPSi layer is removed (figure 5.38f) through chemical etching or DRIE. The second surface of the MPSi layer can then be finished by mechanical polishing.

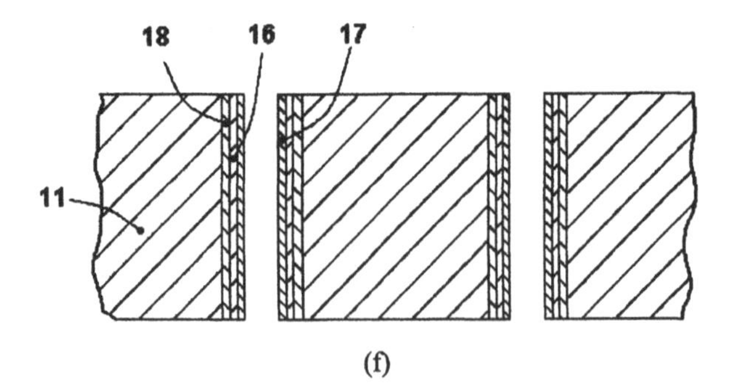

(f)

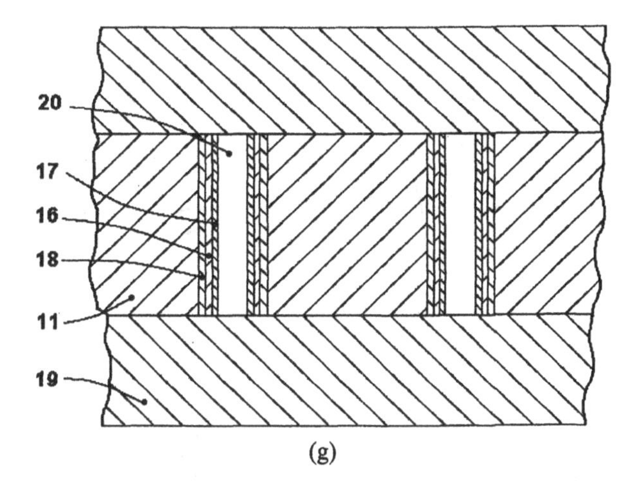

(g)

By following the above fabrication steps, the functional spectral filter will be manufactured. Such a filter can be used in a vacuum or a low-moisture atmosphere. However, in a humid atmosphere, the use of such a filter will be rather limited due to strong absorption of the moisture from the atmosphere through capillary action. Encapsulating the freestanding MPSi layer can solve the problem (figure 5.38g).

5.6 Optical Testing of MPSi-Based Short-Pass Filters

After an MPSi filter fabrication procedure was optimized enough to provide functional filters, MPSi filters were fabricated and extensively tested. This section of this book discusses the results of the optical tests.

First, the filters were tested in Pelkin-Elmer spectrometer in the transmission mode. The distance between the filter surface and the detector exceeded 30 cm, so such measurements can be considered as a far-field measurements. In figure 5.39, the transmission UV to near IR spectra of different MPSi filters are given.

Figure 5.39 presents the experimental normal incidence far-field transmission spectra of the MPSi filters having random-pore structure and different thicknesses of silicon dioxide layer covering the pore walls.

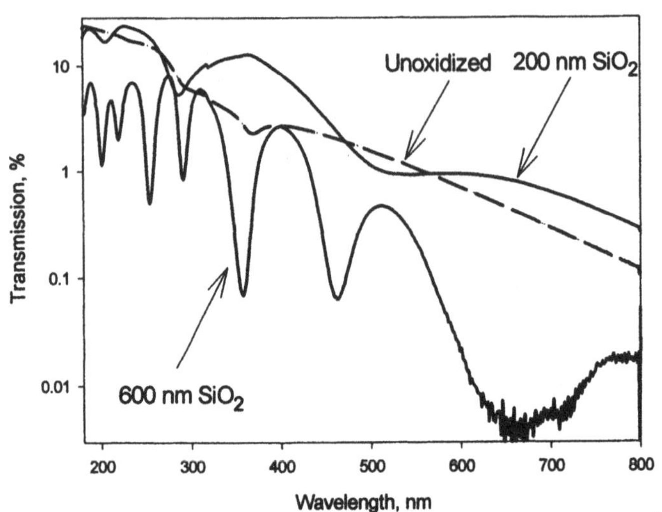

Figure 5.39. Experimental normal-incidence far-field transmission spectra of MPSi-based short-pass filters with different thicknesses of silicon dioxide layer covering the pore walls. Transmission spectra are presented in logarithmic scale.

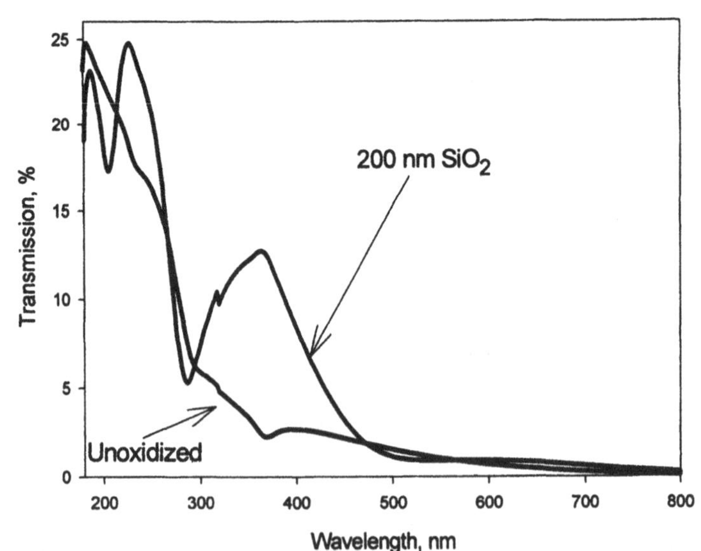

Figure 5.40. Experimental normal-incidence far-field transmission spectra of uncoated and 200 nm SiO_2 coated MPSi-based short-pass filters.

Figure 5.40 gives the experimental normal-incidence far-field transmission spectra of two identically fabricated free-standing MPSi structures, where one of them was post-oxidized so 200 nm of silicon dioxide was formed on the pore walls while another was left untreated. The effect of the additional layer of the transparent material on the transmission spectral shape of the

MPSi short-pass filter is clearly demonstrated. It is also clear from this data that the right oxidation conditions (like temperature and time) have not yet been found to match the shape, predicted in theory (see figure 5.10b). However, the tuning of the transmission spectra by the covering of the pore walls is demonstrated, which validates the theory presented in the beginning of this chapter.

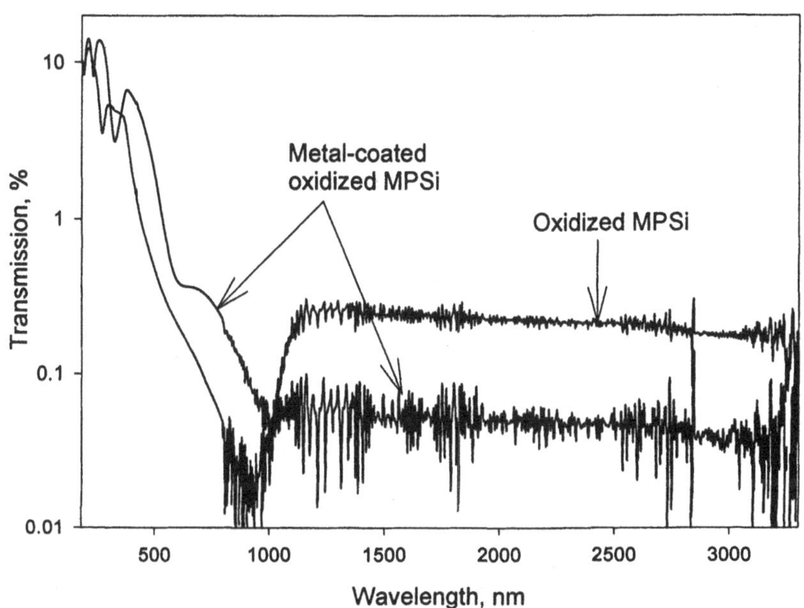

Figure 5.41. Experimental normal-incidence far-field transmission spectra of MPSi-based short-pass filters with and without a metal layer coating one of the surfaces of the MPSi structure.

Another prediction of the theory was the possibility of the suppressing the long-wave tail in the transmittance through the MPSi array by coating at least one interface of a free-standing MPSi array by a layer of metal (see the discussion of equation 5.7). To check it, two quite identical MPSi filters were prepared and one of them was coated by thin (~50 nm) layer of silver in a thermal evaporator. Both these samples were measured in the Perkin-Elmer spectrometer at normal incidence. The experimental transmission spectra of these two samples are presented in figure 5.41. One can see that the metal coating of the MPSi array interface indeed suppresses by an order of magnitude the transmittance at the IR wavelengths, which validates the developed theory. The presence of peaks and valleys caused by the interference in the silicon dioxide layer in the UV and deep UV parts of the transmission spectra for both metal coated and uncoated samples indicates that the thermally evaporated metal was covering mostly the interface of MPSi structure without pore wall coverage, as schematically shown in figure 5.42. Although the suppression of the transmission through a metal-coated

MPSi array was recorded only to 3000 nm wavelength, it is expected that the suppression will extend without limit to the longer wavelengths. Hence, the metal-coated MPSi filter of figures 5.41 and 5.42 is a truly short-pass filter. It should be noted that the suppression of the transmission by two orders of magnitude through the IR spectrum was limited only by the small thickness of the deposited metal layer. The thickness of the deposited metal layer can be increased without any mechanical or thermal stability problems to several hundreds of nanometers to provide the complete suppression of the transmission at the near IR and IR wavelength ranges.

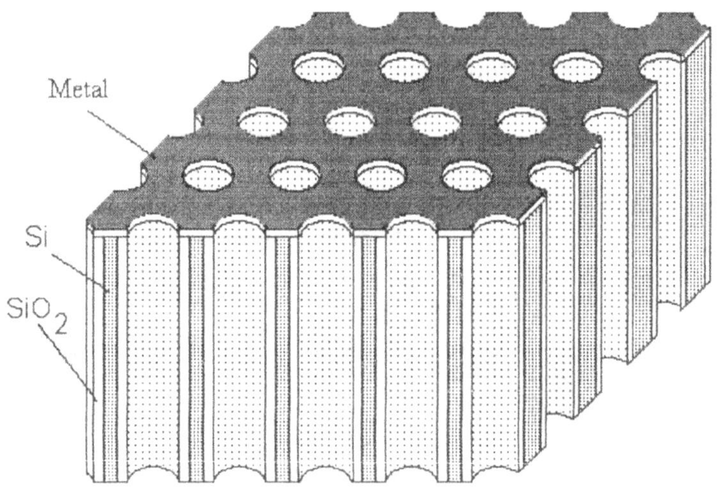

Figure 5.42. MPSi array with uniformly oxidized pore walls and a metal layer covering one of the faces.

The values of far-field transmission of 25% are quite high (in fact, they are about the value of the transmittance through the typical Metal-Dielectric deep UV band-pass filters). However, due to the large number of the diffraction orders for ordered MPSi arrays or due to the scattering for an unordered array, this value is considerably smaller than the value of the near field transmission. In addition, near-field transmission angular measurements were needed to estimate the acceptance angle of the MPSi-based short-pass filter.

The optical setup used for the measurements of the angular dependences of the near-field transmission through the MPSi filters is presented in figure 5.43. The light source used was either frequency doubled Ar+ laser (Coherent, Inc., INNOVA 300 C FRED)) at the wavelength of either 244 or 257 nm or a monochromator (SPEX 1870C Optical Spectrometer, deep-UV/near IR, 0.5 m focal length, 0.01 nm resolution).

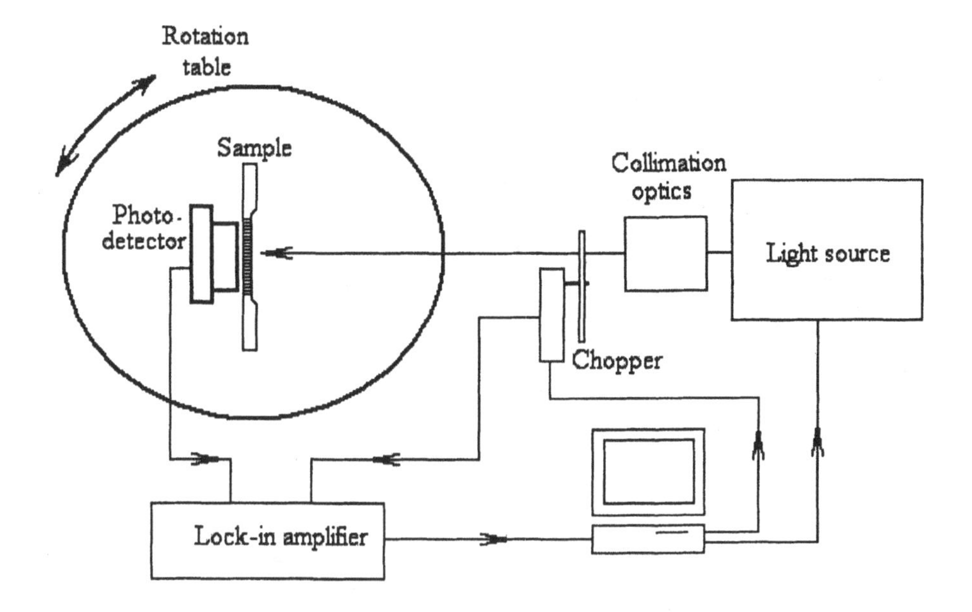

Figure 5.43. Optical setup used for the measurements of the angular dependences of near-field transmission through the MPSi filter.

The collimation optics were used only with the monochromator. An optical chopper and a lock-in amplifier were used to increase the sensitivity and resolution. The sample was pressed to the UV sensitive photodetector and they were placed at the center of an Aerotech ADR 200-11-ND50 computer-controlled rotation stage having 2 arcseconds accuracy over 360 deg. rotation.

The examples of experimentally measured angular dependences of the transmission at 257 nm wavelength through MPSi filters are given in figure 5.44. The measurements were done at approximately 4 mW of CW laser power at 244nm. The photocurrent from the photodetector was below 100 μA, while the saturation current of this particular photodiode was 200 μA. Prior to testing of the MPSi filter, the photodetector was mounted at the center of the rotation stage and its response as a function of the incident angle was used as a reference. Both MPSi filters of figure 5.44 were composed of random macropore arrays (i.e. no initial etch-pits were used). Both samples were obtained from the same silicon wafer with the same electrolyte under the same anodization conditions (current density, anodization time, temperature, etc.). They both were oxidized prior to the backside etching so the silicon dioxide layer (~200 nm) was formed on the macropore walls.

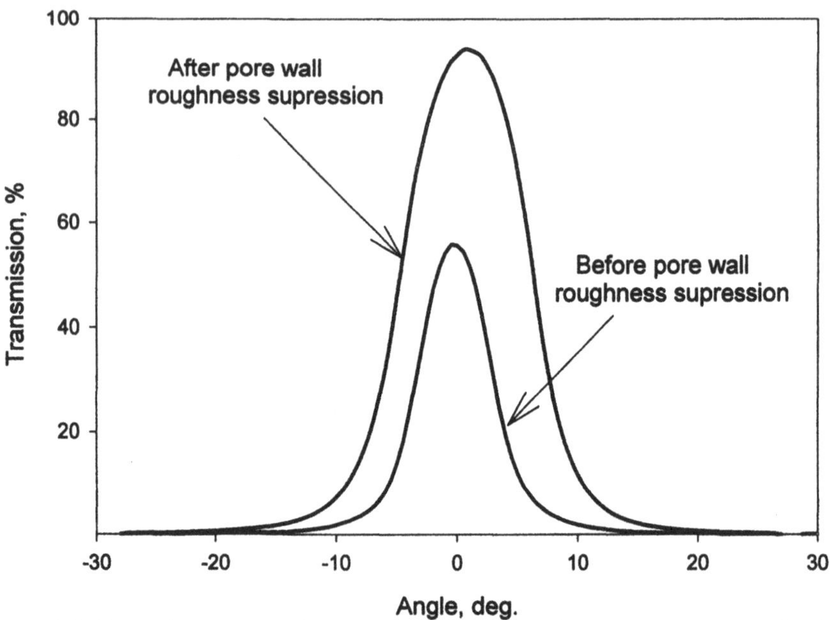

Figure 5.44. Example of the near-field transmission angular dependences of two MPSi filters.

After both filters were completed, one sample was placed into HF solution for 5 minutes under reasonable agitation so all the oxide was removed from the pore walls and the wafer was thermally oxidized again so ~ 200 nm of oxide were re-formed on the pore walls. Such a treatment of the pore walls led to the significant suppression of the pore wall roughness. As a confirmation, in figure 5.45 SEM images of the pore wall crossection before (b) and after (a) pore wall roughness suppression process are shown. While in figure 5.45b the wall roughness is clearly observed, in figure 5.45a the roughness is undetectable even at 2.5 higher SEM resolution. As follows from figure 5.44, such a treatment increased the near-field deep UV transmission from ~ 53% to the value of 94%. Taking into account that the porosity of the samples before the treatment was about 70% and the average pore diameter before the treatment was ~ 3.5 microns, the increase of transmission caused by the increase in porosity was only ~ 11-15%. Another 30% of transmission increase was caused by the increase of the pore wall smoothness, i.e. by the suppression of the Reyleigh scattering (which is proportional to λ^{-4}).

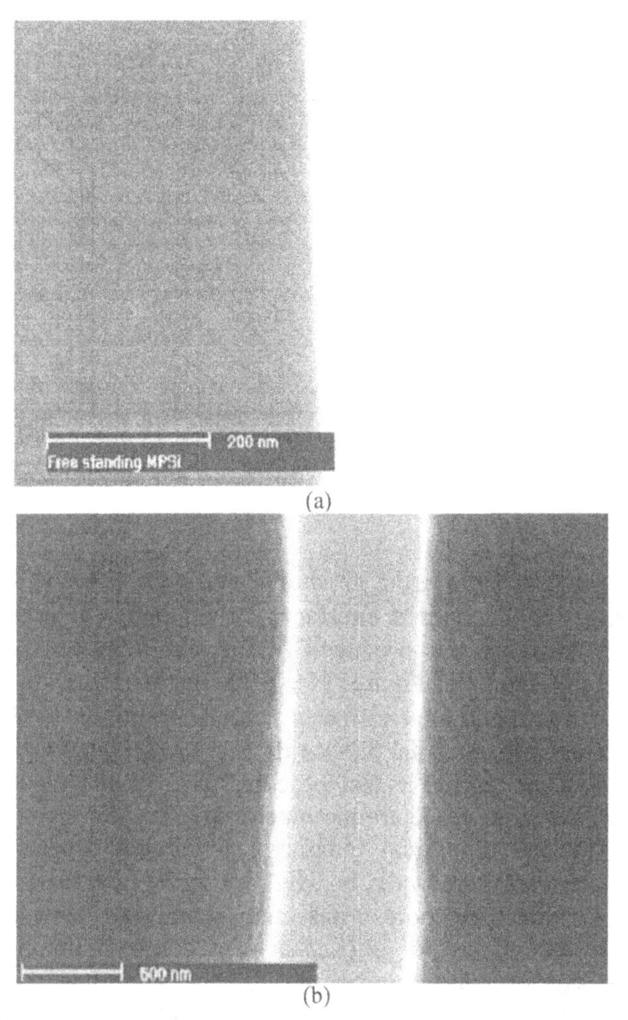

Figure 5.45. SEM images of the pore wall crossection before (b) and after (a) the pore walls roughness suppression.

Even with the suppression of the scattering of the transmitted light on the pore walls, the 94% near-field transmission through the MPSi array requires additional explanation. As was shown in sections 5.2 and 5.3 of this book, the transmission through the MPSi filter cannot exceed the porosity of the MPSi array (see for example equation (5.8)). From front surface measurements it is estimated that the porosity of the MPSi array from the figure 5.44 does not exceeded 75-80%. The explanation of such an extraordinary transmission is based on the fact that the coupling efficiency is proportional to the porosity at the interface of the MPSi structure, which can be quite different from the porosity within the deep part of the MPSi film.

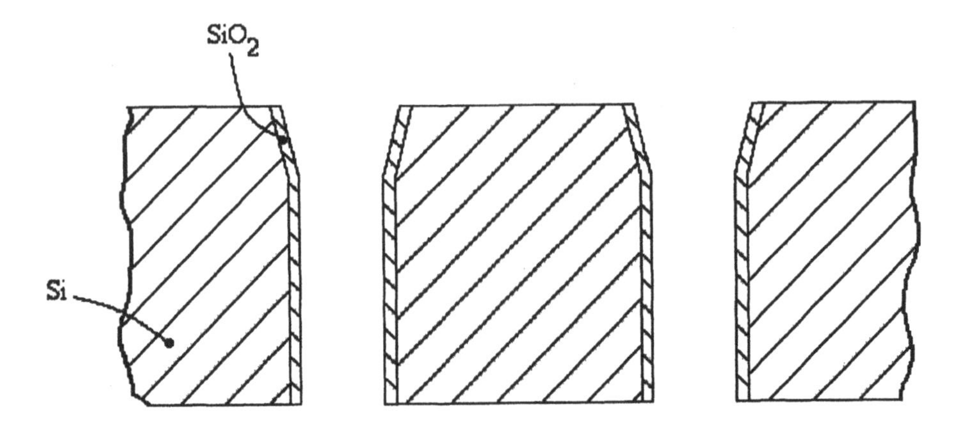

Figure 5.46. Schematic drawing of the MPSi array crossection with the tapered macropore ends.

In fact, tapering of the pore ends, as shown in figure 5.46, will boost the coupling efficiency at the tapered interface, if the tapering is done right (i.e. it is adiabatic). Most of the produced MPSi arrays indeed had tapered interfaces (see, for example, figure 34). This happens due to two main reasons: first, as was mentioned in the process development section of this report, two different processes take place during electro-chemical etching of the silicon: The first relatively fast process involves the electrochemical etching itself, which takes place around macropore tip. The second relatively slow process is the chemical etching of already formed through electrochemical process pore walls by HF containing electrolyte. I.e. while the bottom part of the pores has the diameter, corresponded to electrochemical process conditions (i.e., current density, silicon resistivity, electrolyte composition, etc.), upper part of the pore has the diameter, enlarged with respect to as electrochemically etched by the value, defined by the chemical dissolution of the silicon wall in HF-contained electrolyte. While the second process is slow due to room-temperature of the process and relatively small concentration of HF in the electrolyte, due to long length of the electrochemical process (usually ~ 16 hours) it manifest itself in mentioned tapering.

The near-field transmission spectra of figure 5.44 also tell us about the acceptance angle of porous leaky waveguides. Figure 5.47 gives the Voigt curve fitting of the angular dependence of the near-field transmission through random-pore MPSi filters. It is apparent that Voigt curve fitting provides an excellent fit. The Lorencian part of the Voigt function can be assigned to the broadening of the transmission line due to the contribution from different pore sizes in random MPSi arrays, while the Gaussian part can be assigned to the light emitted from the porous leaky waveguide with an average across-the-beam pore size.

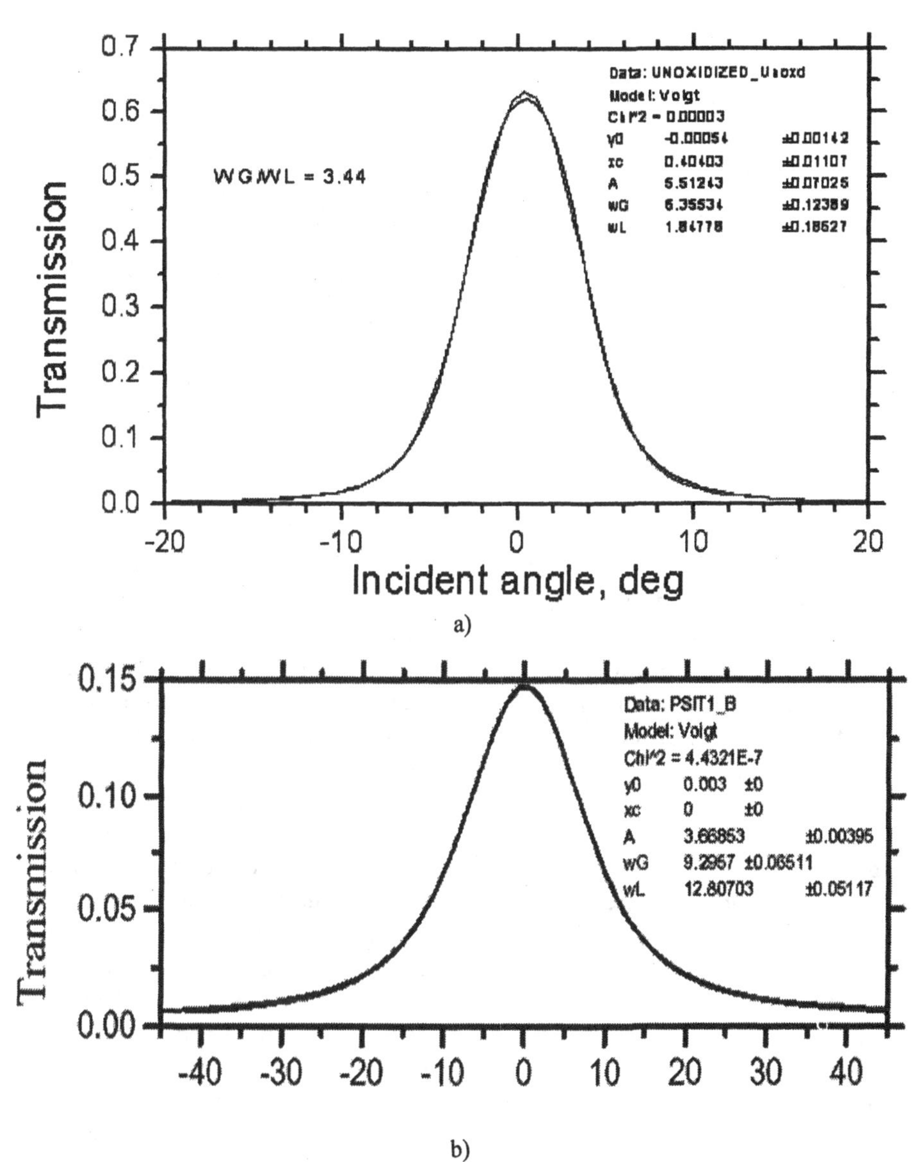

Figure 5.47. Curve-fitting of the angular dependence of the near-field transmission through random-pore MPSi filter.

In figure 5.47b, the curve fitting is given for the angular dependence of the near-field transmission through the random-pore array with the average pore diameter about 1.3 microns. In figure 5.47a, the curve-fitting is given for the angular dependence of the near-field transmission through the random-pore array with the average pore diameter about 3.5 microns. As follows from these plots the acceptance angle of pore leaky waveguide is strongly dependent on the pore diameter. While for the MPSi filter with the average pore diameter about 1.3 microns the acceptance angle range was found to be

(see equation (5.8)). It is expected that by decreasing the pore diameters to 1 micron or below , wide acceptance angle MPSi filters can be produced.

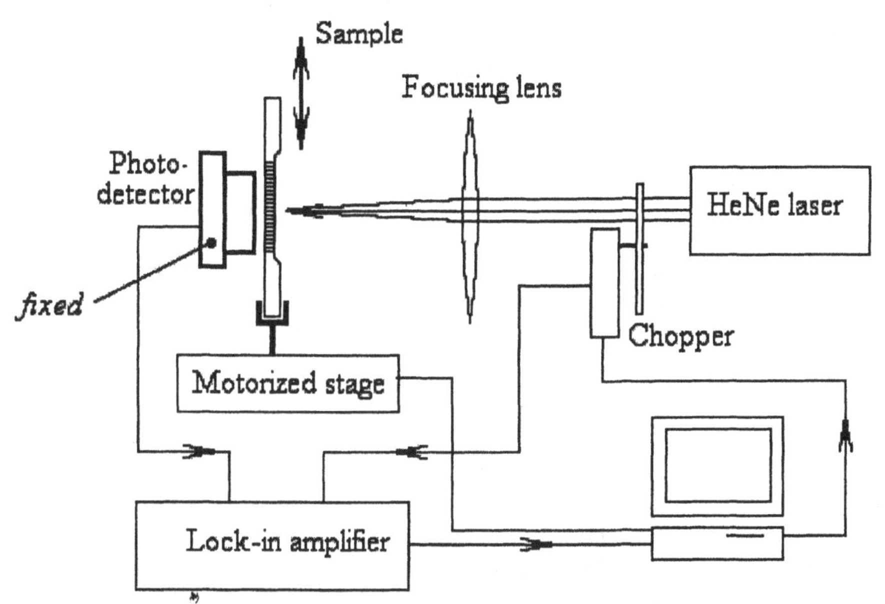

Figure 5.49. Optical setup used for the measurements of the uniformity of the transmission through the MPSi filter.

Next set of experiments was performed in order to check the uniformity of the transmission through the fabricated MPSi filters. The optical setup used for these purposes is shown in figure 5.49. A HeNe laser beam (at 632.8 nm wavelength) was focused by a lens with a focal length of 20 cm onto the surface of the filter. The filter was placed normal to the beam on a single-axis motorized stage. The photodetector was mounted behind the MPSi filter at a distance of 5 mm from filter's second surface on a fixed holder. The motorized stage moved perpendicularly to the laser beam so the beam scanned across the MPSi filter surface. The chopper and lock-in amplifier were used to suppress the background noise and to increase the resolution. Although these MPSi filters were absorptive at red wavelengths (see figures 5.39 and 5.40), the nonabsorbed small portion of the transmitted light was enough for such measurements. It should be noted that the uniformity of the transmission can be scaled from the red wavelength measurements to any other part of the spectrum by using the transmission spectra of figures 5.39 or 5.40 as normalization curves.

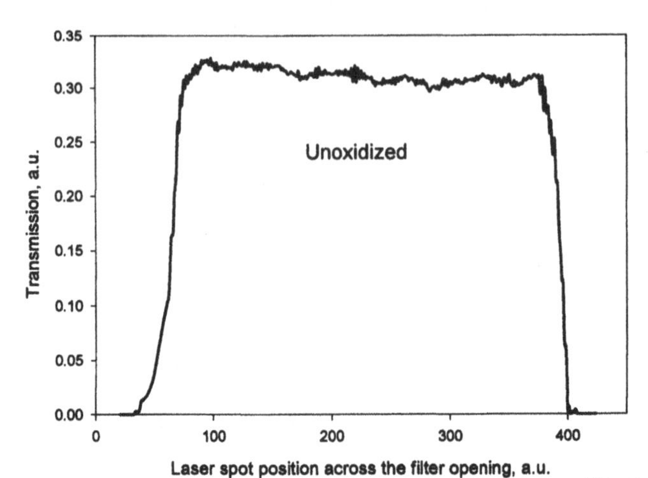

Figure 5.50. Measured transmission profiles across one-inch unoxidized MPSi filter at 632.8nm wavelength.

Figure 5.50a gives the measured transmission profiles across a one-inch unoxidized MPSi filter. The MPSi filter had random-pore structure with average pore diameters ~3.5 microns, porosity of ~70 %, and thickness of ~230 microns. One can see that the uniformity of the transmission through an MPSi filter is quite good (the mean square deviation from average transmission level is within 5%). It clearly indicates that the porosity, average pore diameters and the thickness of fabricated MPSi filters are very uniform across the one-inch free-standing MPSi membrane.

Next set of experiments was performed to check the mechanical stability of the MPSi filters to relatively high-power laser radiation. The filters were exposed to 7 ns 1 J/cm^2 pulses (light from New Age PAL laser was focused by the 20 cm-focal length lens onto the MPSi filter surface normally to it) at 248 nm wavelength for 5 minutes and neither visible damage of the filter structure nor change at the filter transmission were observed. Such amazingly high stability of the MPSi material to the high-power deep UV radiation can be explained as follows: First, the absorption at such wavelength of the MPSi material is quite low (see figure 5.44 – 6% losses there were caused mostly by the coupling losses at the first MPSi interface). Second, these small losses are distributed across all depth of the MPSi material (i.e. 200-200 microns in this case). And third, due to very high porosity and very high surface-to-volume ratio of such a material the silicon-to-air heat exchange is much more efficient than that of common optical materials.

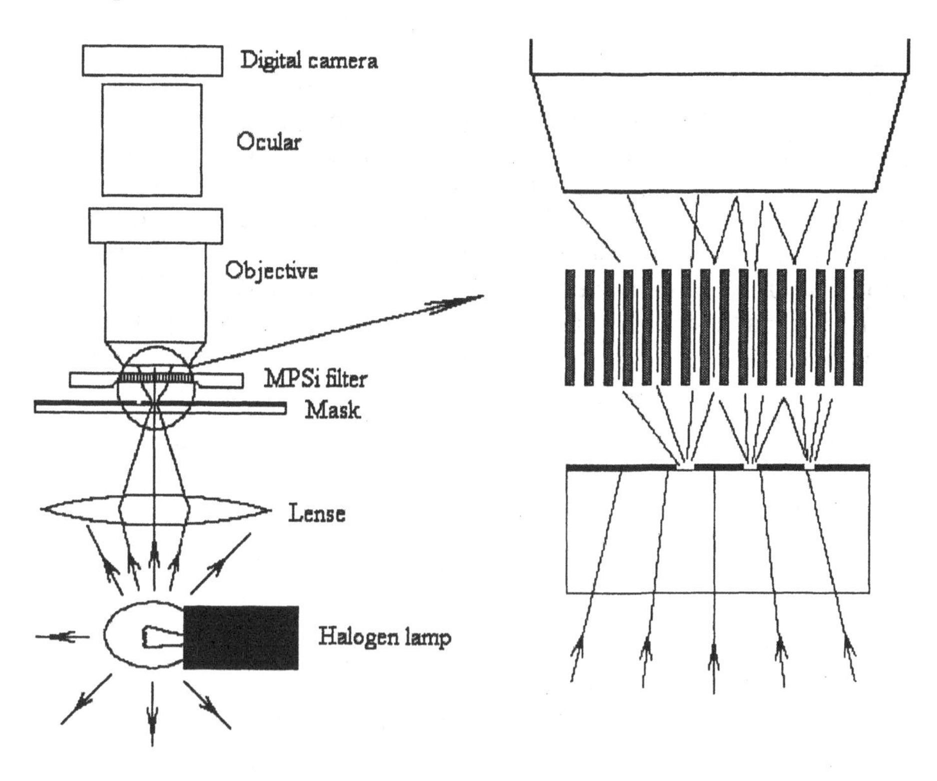

Figure 5.51. Optical setup used for the investigation of the spatial resolution of MPSi filters.

The last set of tests of MPSi filters was performed in order to check the optical resolution limits when an optical image is transmitted though them. The optical setup used for such purposes is given in figure 5.51. The optical microscope was used in transmission mode. The mask, having a pattern with well-known dimensions, was placed into the focus of the microscope and the MPSi filter was placed between the mask and objective. The images were taken by the digital camera, mounted instead of microscope ocular. As expected, random-pore MPSi arrays does not preserve the images transmitted through them. As was explained in the theoretical section of this report, this is caused by the loss of coherence of light transmitted through leaky waveguides having different diameters and, thus, different propagation constants and different phase shifts in transmission. Ordered MPSi arrays, on the other hand, have shown quite good preservation of the image quality. As an example of such images, figure 5.52a shows the image of an alignment mark on the mask without the MPSi filter while figure 5.52b shows the image of the same alignment mark taken through the ordered MPSi filter. This particular image was taken through the square-lattice MPSi array with period 5x5 μm (similar to one shown in figure 5.34). Image features as small as 5 μm (the cross-section of the small cross) were clearly resolved. However, the image contained a number of ghost images caused by higher diffraction orders. A very similar situation was observed with the MPSi array of with 2.5 micron period. The preservation of the resolution of the image when it is transmitted through MPSi filter is a good first result, especially taking into account the complex nature of the transmission of this image(see the right part of figure 5.51). It is possible only due to the excellent preservation of the coherence of porous leaky waveguides across the MPSi filter, which indicates that the quality and uniformity of the macropores were exceptional. The ghost images can be suppressed completely if the MPSi array period can be made on the order of the wavelength of the rejection edge of such filter.

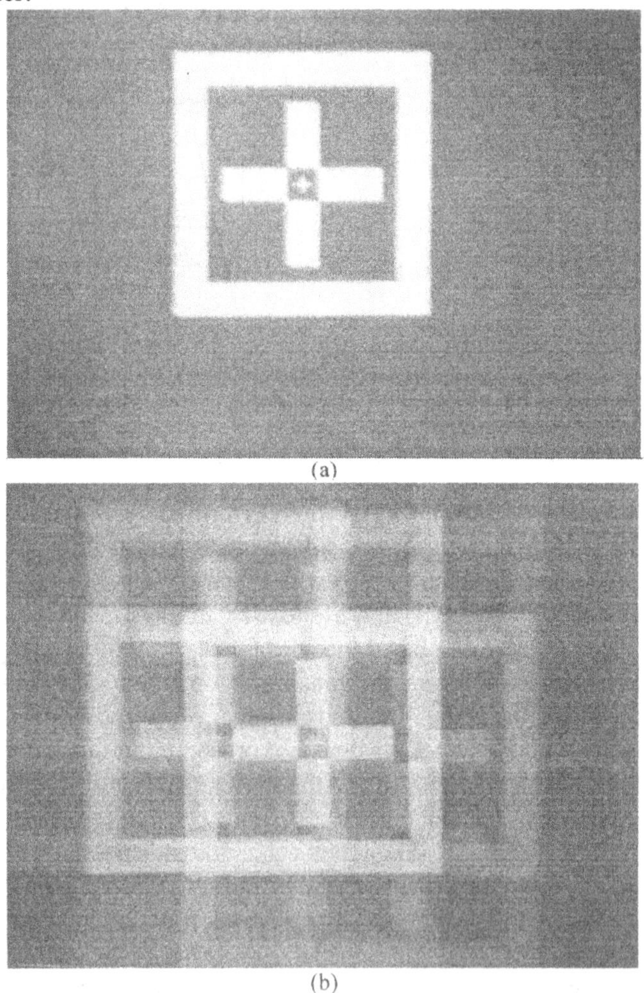

Figure 5.52. Optical microscope transmission image of the alignment mark on the mask (a) and image of the same alignment mark taken through the ordered MPSi filter (b).

Chapter 6

OMNIDIRECTIONAL NARROW-BAND-PASS, BAND-PASS, AND BAND-BLOCKING IR FILTERS

6.1 Introduction

Narrow-band pass, band-pass and band-blocking optical filters are probably among the most frequently used optical devices. Application areas include optical communications, imaging, spectroscopy, lithography, astronomy, and almost any other technology dealing with light. Currently available filters are based on interference in multilayer stacks and have to be used with well-collimated beams and carefully aligned angles of incidence (see section 3.6.2 of this book). Despite adding complexity to optical setup (collimation optics and mechanical alignment), the need for such filters is so strong that they find many applications even with the required additional penalties.

The new approach, similar to one, discussed in Chapter 5 of this book, can be applied for narrow-band-pass, band-pass, and band-blocking optical filter design. Filters designed according to this approach will provide a transmission spectrum substantially independent of the direction of the incident beam (omnidirectional filter) within a much wider angular range than standard filters. The term *omnidirectional* is used here in the loose sense of a wide tolerance and not means that transmittance is 100% over a ±90 deg. range. Such an optical filter has the potential to make a strong impact on optical technology across all sectors of R&D, military, and industrial users.

In this chapter the new design and method of manufacturing will be discussed for such a filter. The design of the new filter is still in the research stage, so the information given here describes the current state of development and is not comprehensive.

6.2. IR Optical Filters, Based on Waveguide Array

In Chapter 5 of this book it was shown that some independent leaky waveguides possess omnidirectional filtering (within the acceptance angles of leaky waveguides). This was achieved due to the independence of the transmission spectral shape of angle of incidence, since the filtering (modification of the transmittance spectral shape) took place during the propagation of the leaky waveguide mode through the leaky waveguides, not during the coupling or outcoupling of light from the leaky waveguide array. The same will be true for the array of independent waveguides. On other hand, it was shown (see section 5.2 of this book) that the macroporous silicon (MPSi) array can be considered in the near IR and IR wavelength as an array of silicon islands waveguides (see figure 6.1). However, the waveguides in the form of silicon islands in MPSi layer are not independent. They are connected with each other by silicon "bridges" and hence will not provide the low level of cross-coupling that is necessary for omnidirectional filtering. Fortunately, relatively simple (from a manufacturing point of view) modifications of the MPSi layer can suppress cross-coupling between silicon island-waveguides. Let us start from the analyses of light transmission through the MPSi layer in the IR wavelength range. The modifications of the MPSi layer, suppressing cross-coupling, are discussed later in the text.

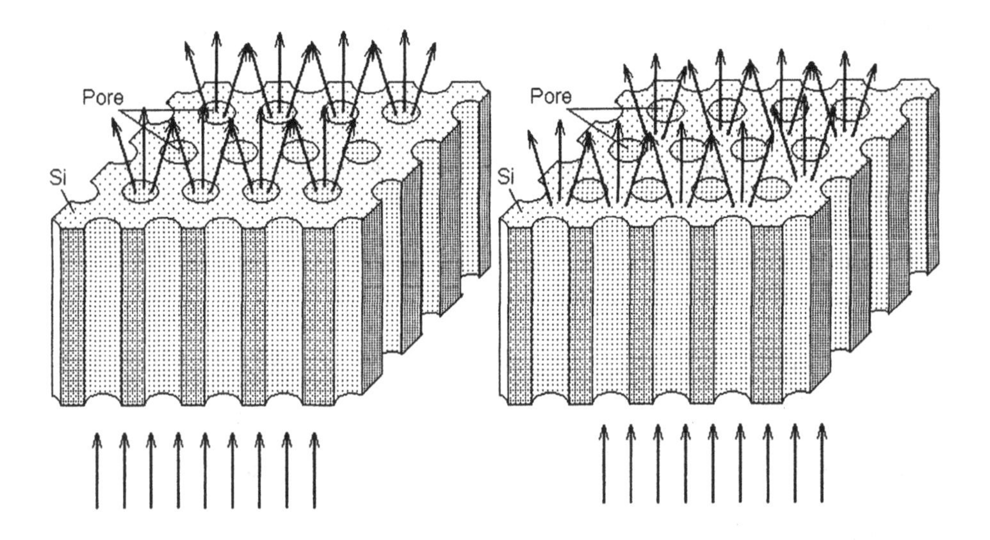

Figure 6.1. Transmission through the MPSi array in the UV visible (left) and IR (right) spectral ranges.

In section 5.3 of this book it was shown that transmission through the MPSi layer have the form:

$$T(\lambda,\theta,\theta') \approx T(\lambda,\theta,\theta') \mid_{\text{deep UV, UV, VIS}}$$

$$+ T(\lambda,\theta,\theta')\mid_{\text{Near IR, IR}} \tag{6.1}$$

where

$$T(\lambda,\theta,\theta')\mid_{\text{deep UV, UV, VIS}} = \\ = \Sigma\, P_i^{LW}(\lambda,\theta)\, DE_i^{LW}(\lambda,\theta')\exp(-\alpha_i^{LW}(\lambda)\, l) \tag{6.2}$$

$$T(\lambda,\theta,\theta')\mid_{\text{Near IR, IR}} = \\ =\Sigma\, DE_i^{W}(\lambda)\, P_i^{W}(\lambda)\exp(-\alpha_i^{W}(\lambda)\, l) \tag{6.3}$$

and where θ is the angle of incidence of light on the first MPSi layer surface (see figure 6.1), θ' is the angle created by the light reemitted from the second surface of the MPSi layer and the second MPSi surface, and other abbreviations are the same as in section 5.2 of this book. Since we are considering near IR and IR spectral ranges, we can neglect leaky waveguide contribution (6.2) with reasonable accuracy. In near IR and IR ranges, unlike UV range, discussed in Chapter 5 of this book, the waveguide losses are negligible from ~1.2 to 7 μm (see figure 6.2) wavelengths (especially taking into account the relatively small thickness of MPSi layer – typically within the 50 to 300 μm range).

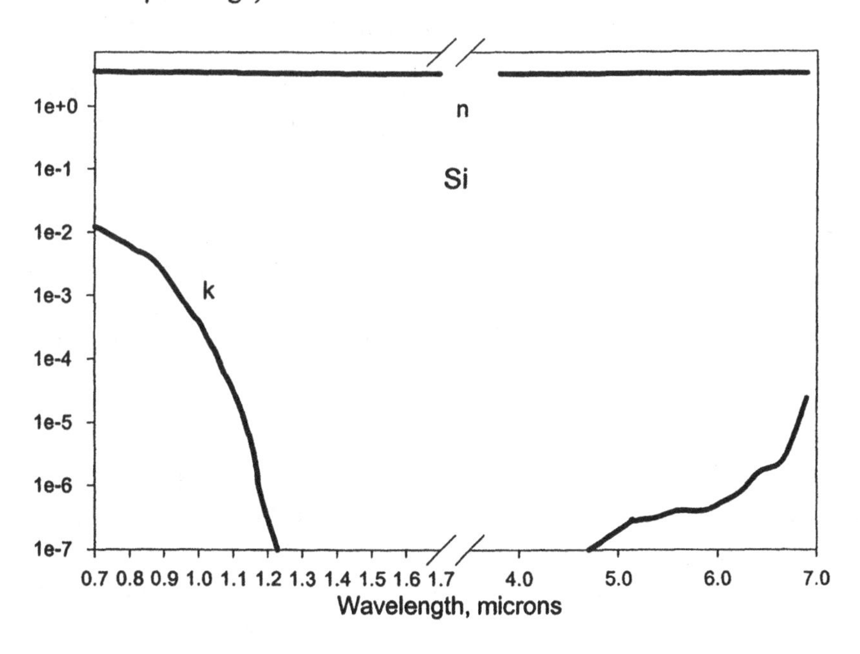

Figure 6.2. Spectral dependences of real and imaginary parts of silicon refractive index in the near IR and IR spectral ranges (after [39]).

Hence, unlike the UV case, waveguide losses cannot provide the transmission spectral characteristics needed for optical filtering applications. However, waveguides are known to exhibit the Bragg reflection phenomenon (see section 2.9 of this book) when their effective refractive indices are periodically modulated in the direction of waveguide mode propagation. Modulation of effective refractive indices in silicon island waveguides in the MPSi layer is possible through modulation of the porosity (or, in other words, modulation of macropore diameters) of the MPSi layer with its depth. Different ways to fabricate such MPSi layers are discussed in the next section. The modulation of MPSi layer porosity will modify the spectral dependence of the transmission through each silicon island waveguide. This modification can be taken into account by introducing an additional term

$$T_i^{BG}(\lambda, \alpha_i^W(\lambda), l')$$

in expression (6.3):

$$T(\lambda,\theta,\theta')\,|_{\text{Near IR, IR}} = \Sigma\, DE_i^W(\lambda,\theta')\, P_i^W(\lambda,\theta)\, T_i^{BG}(\lambda, \alpha_i^W(\lambda), l') \cdot \cdot \exp(-\alpha_i^W(\lambda)\, l) \quad (6.4)$$

where l' is the thickness of the MPSi layer parts that have modulated pore diameters. It is generally less or equal to the thickness of the MPSi layer itself. Since Bragg reflections take place during waveguide mode propagation through the waveguide, under the assumption of the independence (i.e., low cross coupling) of waveguides in the waveguide array, this process is independent on angle of incidence. Hence, if the Bragg grating part $T_i^{BG}(\lambda, \alpha_i^W(\lambda), l')$ of (6.4) contributes most to the overall transmission spectral shape of MPSi layer, this shape will be independent on angle of incidence.

The particular shape of spectral dependence of $T_i^{BG}(\lambda, \alpha_i^W(\lambda), l')$ is a function of silicon islands shape, size, wavelength, waveguide mode losses, and so on (see section 2.9 of this book). It means that the uniform silicon island sizes and shapes are required. The spectral dependence of $T_i^{BG}(\lambda, \alpha_i^W(\lambda), l')$ is different for different orders of waveguide modes. Hence, silicon islands should be single-mode waveguides around the central filtering wavelength. Single mode requirement is also important to ensure independence of transmission spectral shape on the angle of incidence due to different dependences of the coupling and outcoupling efficiencies for different-order waveguide modes. This effect is similar to one discussed for the leaky waveguide case in Chapter 5. However, unlike leaky waveguides, where losses dramatically increase with the number of modes, the losses of waveguide modes are generally small, so in the case of the multimode nature of waveguide, higher-order mode transmission cannot be neglected. To summarize, for the MPSi array to act as an omnidirectional narrowband-pass band-pass or band-blocking filter, it should be ordered, and silicon island waveguides should be independent and single-mode. In this case expression (6.4) will take the form

$$T(\lambda,\theta) = \\ = DE_0^{W}(\lambda,\theta)\, P_0^{W}(\lambda,\theta)\, T_0^{BG}(\lambda, \alpha_0^{W}(\lambda), l')\, \exp(-\alpha_0^{W}(\lambda)\, l) \qquad (6.5)$$

since for the ordered array of waveguides $\theta = \theta'$(see the discussion given in section 5.3).

The coupling efficiency at the first MPSi layer surface can be estimated as

$$P_0^{W}(\lambda,\theta) = P_0^{W}(\lambda,\theta=0)\exp(-(\theta/\theta_{ac})^2) \qquad (6.6)$$

where $P_0^{W}(\lambda,\theta=0)$ is the coupling efficiency in the waveguide array of plane-parallel beam incident normally to the MPSi layer, and θ_{ac} is the acceptance angle of the silicon island waveguide. As was shown in section 5.2 of this book, $P_0^{W}(\lambda,\theta=0)$ can be estimated as

$$P_0^{W}(\lambda,\theta=0) \approx \frac{4n_{Si}(\lambda)\cdot n_I}{(n_{Si}(\lambda)+n_I)^2}\cdot\frac{S_{uc}-S_p}{S_{uc}} \qquad (6.7)$$

where $n_{Si}(\lambda)$ is the refractive index of the silicon at wavelength λ, S_p is the area of each pore, S_{uc} is the area of a MPSi array's unit cell, and n_I is the refractive index of the incident medium. For the most common case of incidence from air (6.7) can be rewritten as

$$P_0^{W}(\lambda,\theta=0) \approx \frac{4n_{Si}(\lambda)}{(n_{Si}(\lambda)+1)^2}\cdot\frac{S_{uc}-S_p}{S_{uc}} \qquad (6.7a)$$

In other words, to some approximation, $P_0^W(\lambda,\theta=0) \approx 0.69(1 - p(0))$, where $p(0)$ is the porosity of the MPSi layer near the first MPSi layer interface. It should be noted, that the porosity (i.e., macropores diameters) is not constant across the MPSi layer. As was discussed above, for narrow-band-pass, band-pass, or band-blocking filter applications, MPSi porosity has to be modified in a periodic fashion through the MPSi layer depth that is

$$p(x) = p_0 \sin(2\pi x/\Lambda + \phi),\ x \in [x_0, x_0 + l'].$$

Moreover, the porosity at both surfaces of the MPSi layer can be minimized within some limits to maximize the coupling and outcoupling efficiencies of the silicon islands waveguide array. Hence, (6.6) can be rewritten as

$$P_0^W(\lambda,\theta) = 0.69(1 - p(0)) \exp(-(\theta/\theta_{ac})^2) \qquad (6.8)$$

As was discussed in section 2.6.1 of this book, the particular value of the acceptance angle strongly depends on waveguide structure. For silicon island waveguides and air-filled pores, the $\sin\theta_{ac}$ estimated according to (2.60) will be $[(n_{Si})^2-1]^{1/2} = 1.6$ that is acceptance angle of silicon island waveguide (and through that MPSi layer as a waveguide array) will be $\pi/2$. Such a wide acceptance range of the MPSi layer at IR wavelength range means that spectral filters based on the MPSi layer will be really omnidirectional (unlike the deep UV and UV cases, where acceptance angles were around 8 to 14 degrees).

0th order diffraction efficiency, $DE_0^W(\lambda,\theta)$, is expected to behave like for the leaky waveguide array discussed in section 5.3. It will depend on porosity $p(l)$ at the second interface of the MPSi layer, on the wavelength of light, on the geometry of the MPSi array, and so on.

We have already defined the properties that the MPSi layer should demonstrate, so let us discuss how it can be implemented. First (and the most important property for ensuring omnidirectional transmission) is the cross-talk suppression between neighbor silicon island waveguides. The simplest way to suppress it with respect to the pure silicon MPSi layer is to partially oxidized the pores walls, since the refractive index of silicon dioxide (1.44 at 1550 nm wavelength) is considerably smaller than that of the silicon (3.5 at 1550 nm wavelength). The schematic drawing of such an MPSi array is given in figure 6.3.

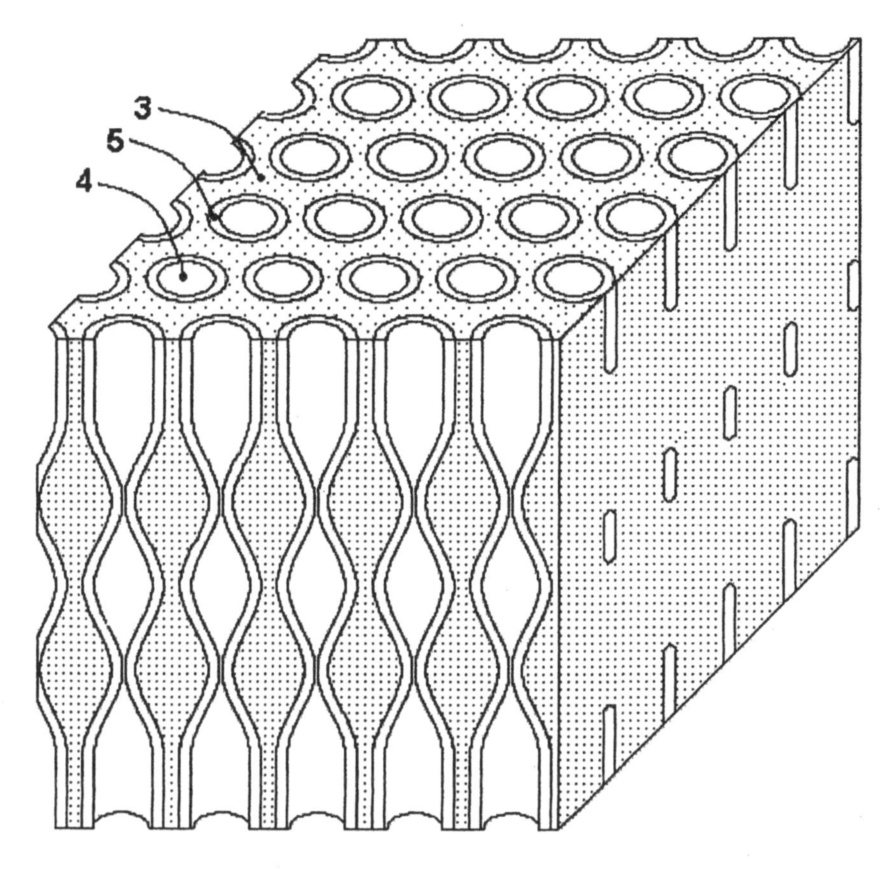

Figure 6.3. An MPSi array with coherently modulated pore diameters and partially oxidized macropore walls.

The MPSi layer will consist of air- or vacuum-filled macropores (4), having coherently modulated pores diameters at least on part of the MPSi layer thickness, thin layer of thermally grown silicon dioxide (5), uniformly covering macropores walls, and silicon islands (3).

As with short-pass filter applications, discussed in Chapter 5 of this book, there is a big variety of MPSi array symmetries that can be implemented. As an example, in figure 6.4 three types of MPSi array cross-section are shown: (a) cubic symmetry, near-square pores; (b) cubic symmetry, circular pores; (c) hexagonal symmetry, circular pores. The unit cells are denoted as (7) and expected waveguide mode profiles are denoted as (6).

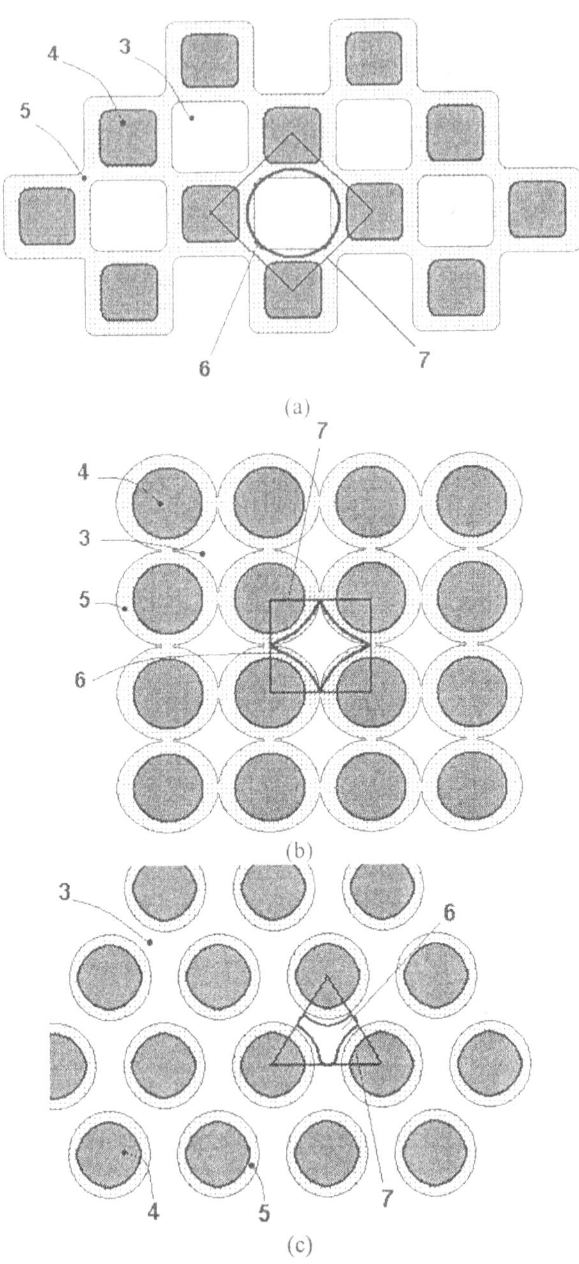

Figure 6.4. Possible symmetries of an MPSi array: (a) cubic symmetry, near-square pores; (b) cubic symmetry, circular pores; (c) hexagonal symmetry, circular pores.

From this drawing we can conclude that hexagonal-symmetry array is disadvantageous from the viewpoint of combining low cross-coupling and reasonable coupling efficiency at the first MPSi interface (ratio of area of waveguide mode to the area of array unit cell). By comparing figures 6.4a and 6.4b we can conclude that cubic MPSi array with near-square pores looks advantageous with respect to cubic MPSi array with circular pores. For the case of figure 6.4a the coupling efficiency at the first MPSi interface is expected to exceed 50%.

Before estimating cross-coupling between neighbor waveguides, let us first find the silicon island size and silicon dioxide layer thickness ranges needed to guarantee the single-mode character of the silicon island waveguides.

In figure 6.5 the numerically calculated dependences of effective refractive indices of TE polarization waveguide modes on the silicon island cross-section are presented for the cubic-symmetry array of near square pores. The wavelength of light was assumed to be 1550 nm. One can see that the silicon island waveguides became multimode starting from about 300 nm of silicon island cross-section. For other wavelengths the maximum silicon island cross-section needed can be found by scaling.

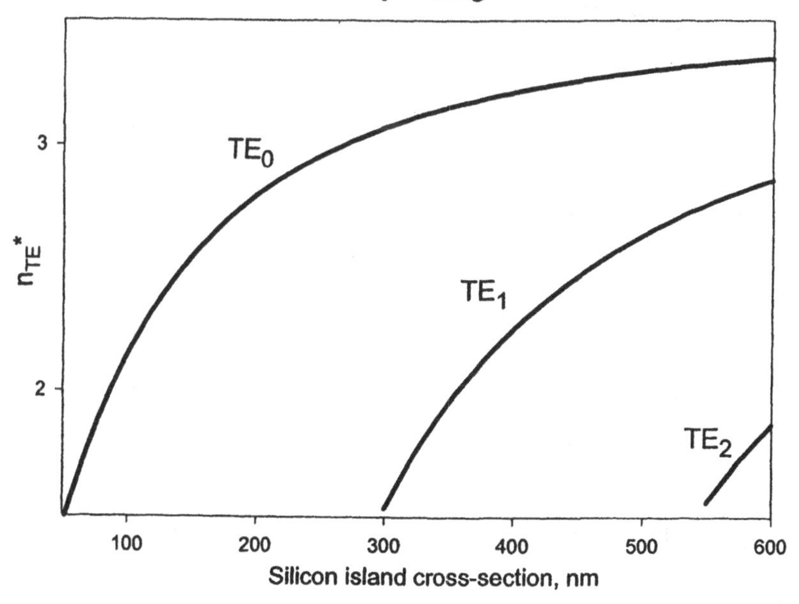

Figure 6.5. Numerically calculated dependences of effective refractive indices of TE polarization waveguide modes on the silicon island cross-section.

Silicon dioxide layer thickness affects both the cross-coupling suppression and the number of modes in the silicon island waveguide for constant silicon island cross-section. In figure 6.6 the numerically calculated dependences of effective refractive indices of TE polarization waveguide modes on the silicon dioxide layer thickness are given for the MPSi layer of figure 6.4a. The wavelength of light was assumed to be 1550 nm, while the silicon island cross-section was assumed to be 275 nm. One can see that even for such a silicon island cross-section the thickness of the silicon dioxide layer covering pore walls has to be less than approximately 300 nm. Now having the silicon island size, the silicon dioxide layer thickness, and the geometry of the MPSi array, we can estimate the cross-coupling coefficient between neighbor waveguides. The method used for such purposes was discussed in section 2.10 of this book. In figure 6.7 the numerically calculated dependence of the cross-coupling coefficient on the silicon island separation is given for fundamental waveguide modes. The wavelength was assumed to be 1550 nm, the silicon island cross-section was assumed to be 275 nm, the silicon dioxide layer thickness was assumed to be 300 nm, and the MPSi layer structure was assumed to be the same as in figure 6.4a.

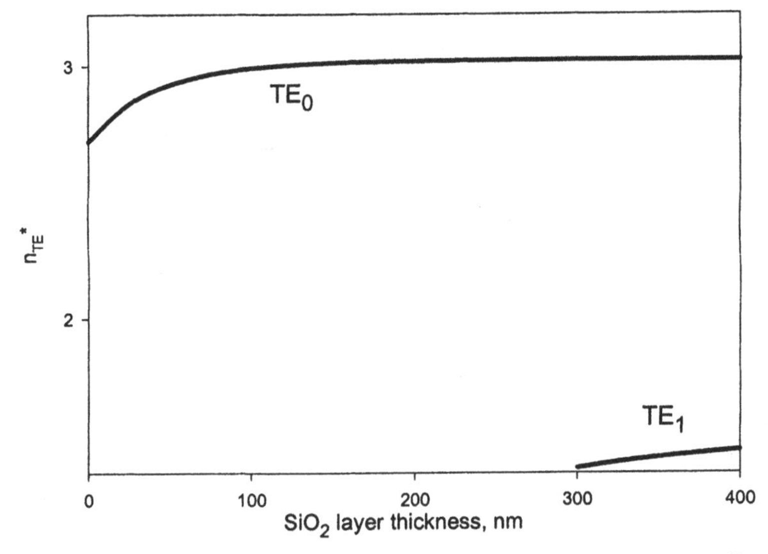

Figure 6.6. Numerically calculated dependences of effective refractive indices of TE polarization waveguide modes on the silicon dioxide layer thickness for the structure of figure 6.4a.

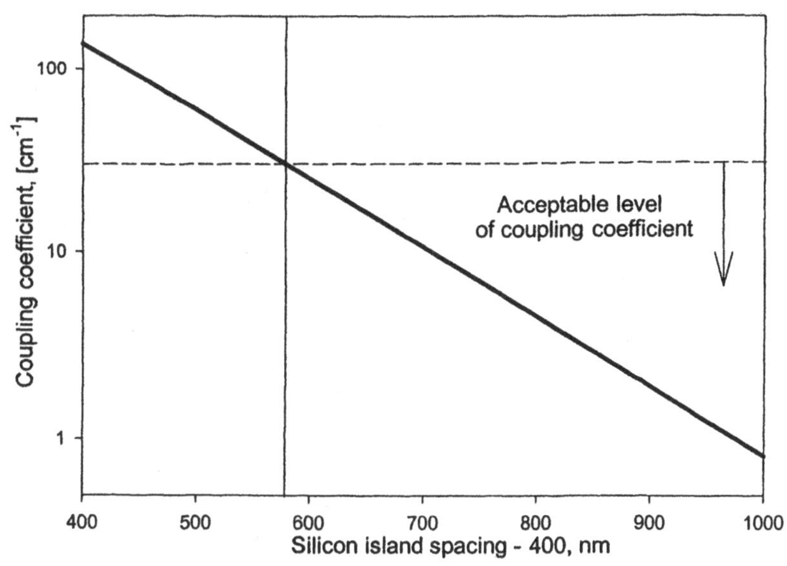

Figure 6.7. Numerically calculated dependences of cross-coupling coefficients of fundamental waveguide modes on the silicon island separation.

As expected, the coupling coefficient decreases exponentially as the silicon island spacing decreases. To provide omnidirectionality of the transmitted light, the coupling length, which is equal to the inverse coupling coefficient, should be less than the thickness of the MPSi layer. Taking into account that the typical MPSi layer thickness is in the range of 50 to 300 μm, while the thickness of the MPSi layer with modulated pores diameters will not exceed 200 μm, it will be safe to set the upper limit of the coupling coefficient at about 20 cm^{-1}.

According to figure 6.7, the cross-coupling coefficient will be suitable for providing omnidirectionality of the transmittance spectrum for silicon island spacing starting from ~980 nm. The MPSi array is assumed to be of cubic symmetry; hence, the unit cell area in this case will be about 1 μm^2 (see figure 6.4a). The waveguide mode area will be 0.16 mm^2 for the 1550 nm wavelength. Hence, the coupling losses at the first MPSi layer interface (if pores and hence silicon islands are constant across all MPSi depth) will be around 84%. Such a value is absolutely not suitable for any optical filter applications. Even modification of the porosity of the MPSi layer near both interfaces will not boost the potential transmittance to more than 25 to 35%. Hence, a better way of suppressing the cross-coupling while keeping coupling losses at a reasonable level is needed. It is possible by making further modifications in MPSi layer structure.

One relatively simple way to modify the MPSi layer structure to suppress cross-coupling while keeping the potential transmittance at a reasonable level is to fill the pores with highly reflective material, such as metal. A thin layer of low refractive index material (for example, thermally grown silicon dioxide) is still needed between the metal, filling the pores, and silicon island waveguides to suppress the propagation losses in the waveguides. The schematic drawing of such an MPSi array is given in figure 6.8.

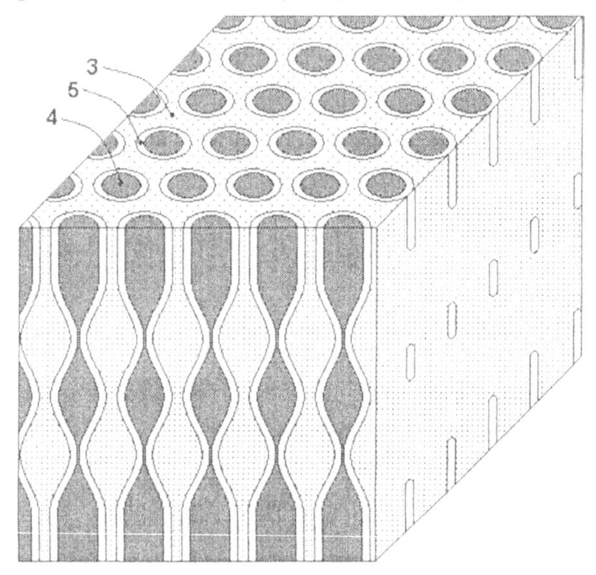

Figure 6.8. An MPSi array with coherently modulated pore diameters, partially oxidized macropore walls, and pores filled with reflective material.

The structure of such an MPSi layer is very similar to one discussed previously (compare figures 6.8 and 6.3). The only difference is that the pores in this case are filled by the metal rather than by the air. Hence, the arguments about preferred MPSi array geometry given in the discussion of figure 6.4 can be transferred to the present case: we will still consider the MPSi array of cubic symmetry and near-square macropores.

Before investigating the particular values of coupling efficiencies and cross-coupling coefficients, we need to define the sizes of silicon islands and the thickness of the silicon dioxide layer needed to maintain single-mode operations of silicon island waveguides and the metal-caused propagation losses on a sufficiently low level.

The silicon island cross-section can be taken from the previously given consideration (see figure 6.5). Hence, we can set the maximum silicon island cross-section to be around 300 nanometers to ensure single-mode waveguiding. However, we need to pay additional attention to propagation losses, absent in previous case but considerable here. As an example, in figure 6.9 the numerically calculated dependences of loss coefficients of TE-polarized waveguide modes on the silicon island cross-section are presented. The MPSi layer structure was assumed to be of cubic symmetry; the silicon dioxide layer thickness was assumed to be 200 nm, and calculations were made for 1550 nm wavelength. In the calculations the metal was assumed to be nickel.

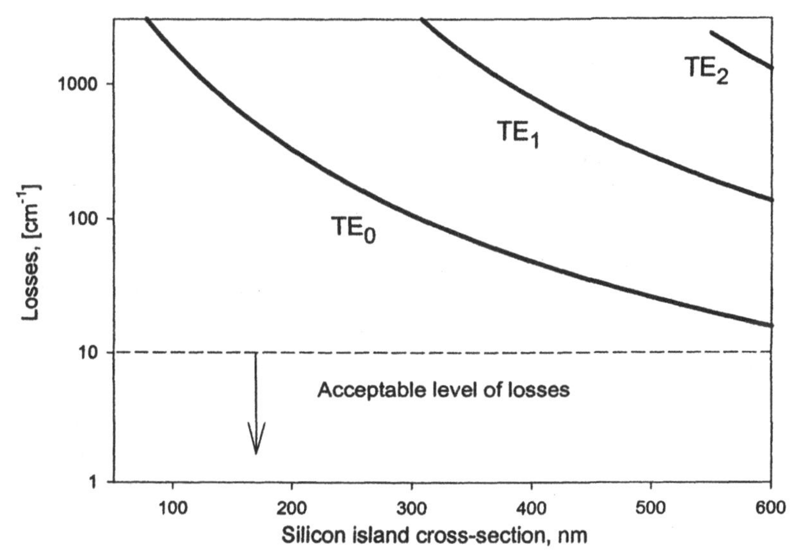

Figure 6.9. Numerically calculated dependences of loss coefficients of TE polarization waveguide modes on the silicon island cross-section.

The waveguide mode losses decrease for each mode with the increase of the silicon island cross-section. It happens due to the increase of the localization of the waveguide mode in the silicon island and, hence, the decrease of the localization of the waveguide mode near the metal/silicon dioxide boundary when the waveguide parameters are getting further from the cut-off conditions. The important question here is the acceptable level of losses that are sufficient to provide a reasonable performance of narrow-band-pass, band-pass, or band-blocking filter, based on such an MPSi layer. From one point of view, the coupling losses at the first MPSi layer interface will be dominating (probably not less than 20 to 30%): even the losses of 10% of coupled intensity due to propagation losses will not cause the dramatic change of the performance of the MPSi-based filter. Taking into account that MPSi layer thickness will not exceed 300 μm, it will give maximum acceptable propagation losses of around 10 cm^{-1}. From another point of view, the quality of waveguide Bragg resonance is the function of the level of losses in waveguide: high losses will be prohibitive in obtaining a really narrow pass band (in the case of narrow-band-pass filter) or steep edges of pass-band (in the case of band-pass or band-blocking filters). However, these filter parameters are unique for each filter design, as is the acceptable level of waveguide mode propagation loss. Hence, we can set the level of maximum acceptable losses at 10 cm^{-1} but note that this value can be significantly less for specific filter designs.

It is obvious that for the MPSi structure of figure 6.9 the losses are way too high. However, the losses are known to drop exponentially with the buffer (silicon dioxide in this case) thickness. As an example, figure 6.10 gives the numerically calculated dependences of loss coefficients of TE polarization waveguide modes on the silicon dioxide layer thickness. The MPSi layer structure was assumed to be of cubic symmetry; the silicon island cross-section was assumed to be 275 nm, and calculations were made for 1550 nm wavelength. In the calculations the metal was assumed to be nickel. In this particular example the losses became acceptable for a silicon dioxide layer thick enough to cause the appearance of a second waveguide mode. However, unlike the MPSi layer with pores not filled with metal, it will not cause problems due to very high losses of this mode (two orders higher than that of fundamental mode). Hence, we can conclude that the MPSi layer structure should be as follows: the silicon island cross-section should be not more than 300 nm, and silicon dioxide thickness should be not less than 300 nm.

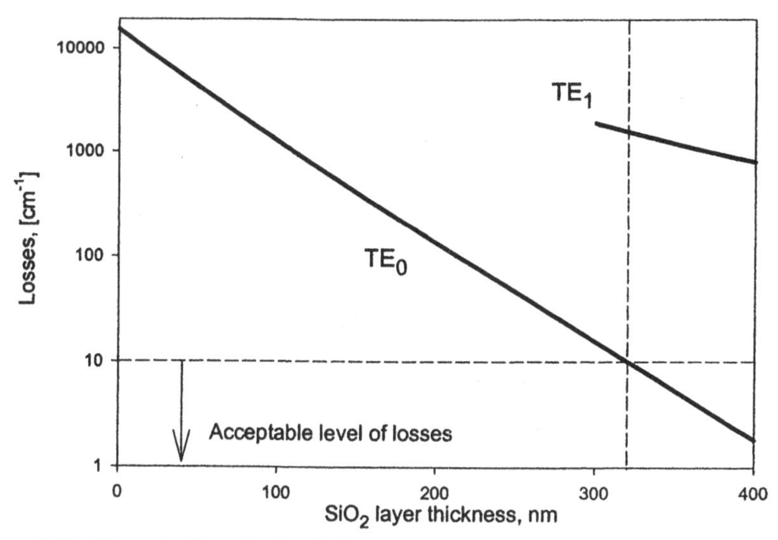

Figure 6.10. Numerically calculated dependences of loss coefficients of TE polarization waveguide modes on the silicon dioxide layer thickness.

With these parameters in mind, we are now ready to estimate the level of cross-coupling and get the approximate sizes of MPSi array. In figure 6.11 the numerically calculated dependence of cross-coupling coefficient on the silicon island separation is given for the fundamental waveguide mode. The wavelength was assumed to be 1550 nm, the silicon island cross-section was assumed to be 275 nm, the silicon dioxide layer thickness was assumed to be 300 nm, and the MPSi layer structure was assumed to be the same as in figure 6.4a with pores filled with nickel. By comparing figures 6.7 and 6.11 one can see that metal in the pores indeed suppress the cross-coupling between neighbor silicon island waveguides. Cross-coupling suppression is not very strong. However, it should provide better potential coupling efficiency at the first MPSi layer interface due to a smaller acceptable value of porosity. We can estimate it the same as in the air-filled MPSi layer case: according to figure 6.11, the minimum acceptable (from the viewpoint of the cross-coupling suppression) silicon island spacing of 910 nm corresponds to the area of such an MPSi array unit cell of ~ 0.82 μm^2. The waveguide mode area will be 0.2 μm^2 for the 1550 nm wavelength. Hence, the coupling losses at the first MPSi layer interface, if pores and hence silicon islands have constant sizes across all MPSi depth, will be ~75%. With the modification of the porosity of the MPSi layer near both interfaces the potential transmittance can be increased to about 40 to 50%.

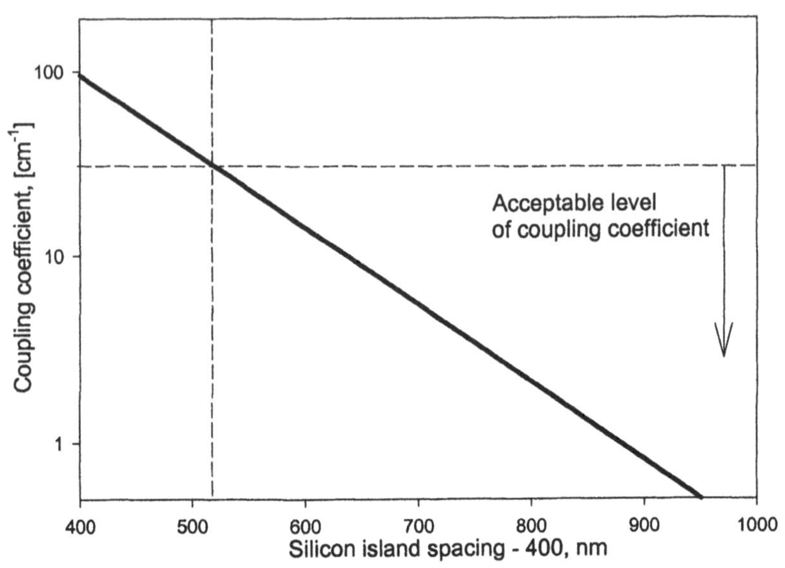

Figure 6.11. Numerically calculated dependences of cross-coupling coefficients of fundamental waveguide modes on the silicon islands separation for metal-filled pores.

This value of potential transmittance is small, but it is higher than that for air-filled pores. The appearance of metal-caused propagation losses in this design is negative from the viewpoint of quality of Bragg resonance. However, since the value of propagation losses is very different for different orders of waveguide modes, it is possible to utilize two-mode waveguides in such an MPSi layer, since a second-order mode will dissipate due to high losses. It provides an opportunity to utilize higher than 300 nm (for 1550 nm wavelength) cross-sections of silicon island waveguides for the same area of an MPSi array unit cell (i.e., it increases the coupling efficiency at the first interface of MPSi layer).

The numerically calculated spectral dependences of the coupling, outcoupling, and propagation losses of a nickel-filled cubic MPSi array with period (pore-to-pore distance of 1500 nm) pore diameters (after oxide layer growth step) of 750 nm and a 300 nm oxide layer covering the pore walls is presented in figure 6.12. One can see that propagation losses are expected to be considerably less than coupling losses for such a filter.

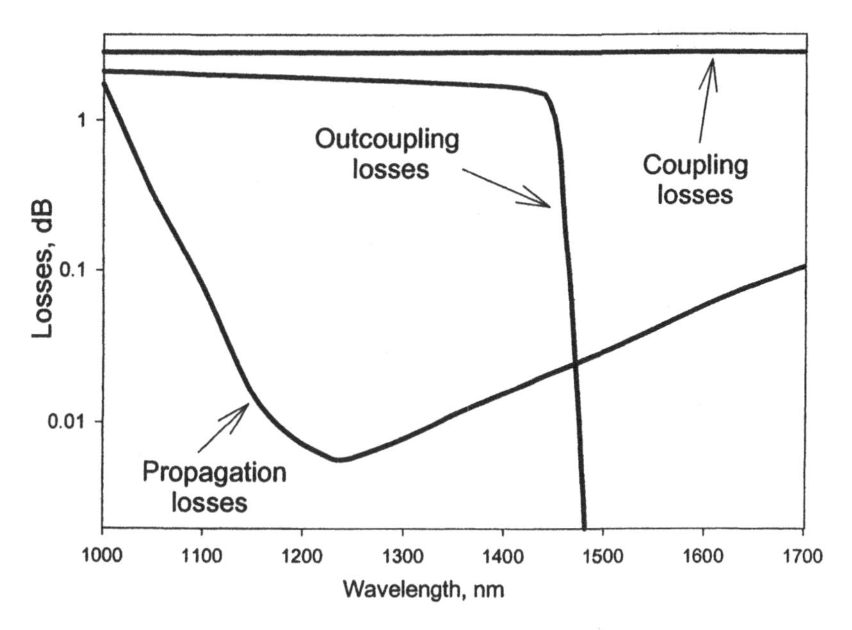

Figure 6.12. Numerically calculated coupling, propagation, and outcoupling losses for the structure with parameters given in the text.

The metal-filled MPSi array of figure 6.12 clearly can find some applications, since despite a considerably low level of transmittance, it offers for narrow-band-pass, band-pass, or band-blocking filters an independence of transmittance spectral shape on angle of incidence. However, further suppression of losses can expand the potential markets for such a filter.

Another solution of this tradeoff between cross-coupling and coupling losses is to create MPSi arrays of more complex symmetries. The examples of such MPSi array cross-sections are given in figure 6.13. In figure 6.13a the advanced hexagonal symmetry MPSi array is presented, while in figure 6.13b the advanced cubic symmetry MPSi array is presented. In both cases the macropores are of circular cross-section. Both the array unit cells and the silicon island waveguide mode profile are schematically shown. One can see that for MPSi layers in figure 6.13 the ratio of silicon island waveguide mode to MPSi array unit cell is considerably higher than that of the MPSi layers of figure 6.4. It should be noted that macropores in advanced-symmetry MPSi structures could be also filled either with air or with metal (if more suppression of cross-coupling is required).

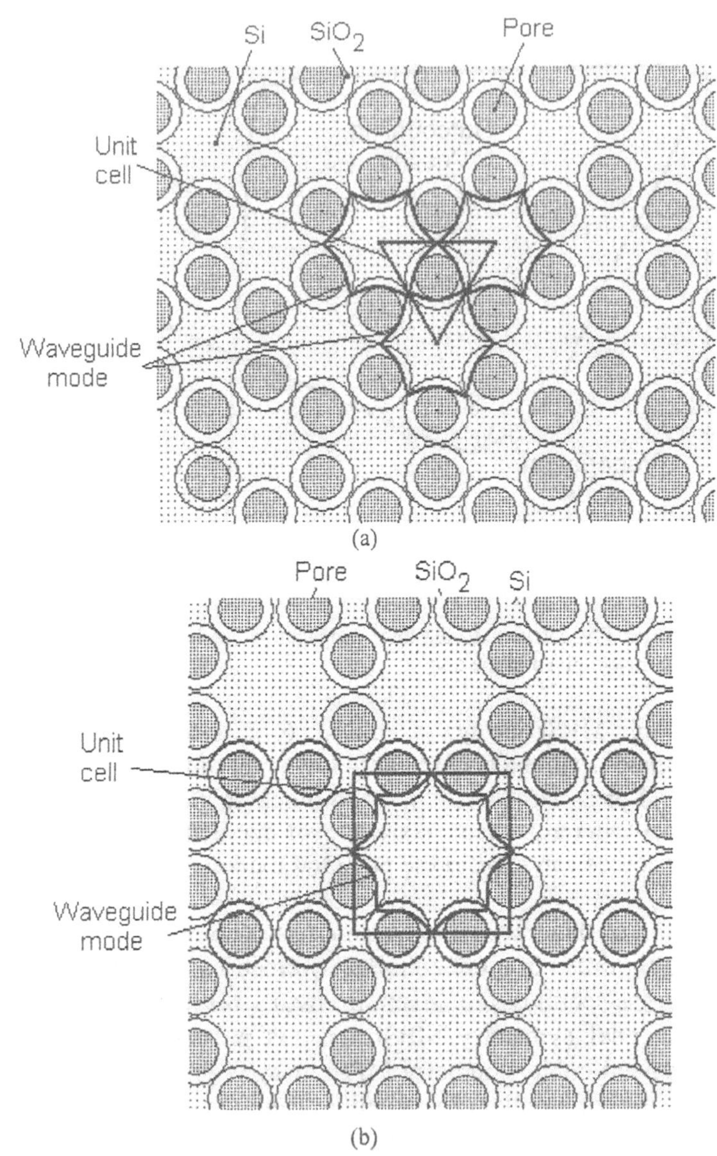

Figure 6.13. Possible symmetries of an MPSi array: (a) advanced hexagonal symmetry; (b) advanced cubic symmetry. In both cases the macropores are of circular cross-section.

The required dimensions of the MPSi layer can be transferred at some extent from the previous discussion. The cross-section of silicon islands should be less than 300 nm for 1550 nm wavelength. The size of macropores and the thickness of the silicon dioxide layer covering the pore walls should be found according to cross-coupling suppression requirement. In figure 6.14

the numerically calculated dependences of cross-coupling coefficients of fundamental waveguide modes on the silicon islands separation are given for the MPSi layer of figure 6.13a. The wavelength was assumed to be 1550 nm, the pores were assumed to be circular, and the thickness of the silicon dioxide layer was assumed to be equal to halve of the macropore diameter. As follows from figure 6.14 the cross-coupling coefficient between neighbor silicon island waveguide fundamental modes reaches acceptable levels at the silicon island separation of 790 nm.

By taking simple geometrical calculations and taking into account the single-mode requirement we can find that the needed macropore diameter should be around 180 nm and the needed thickness of the silicon dioxide layer should be at least 120 nm. The unit cell area for an MPSi layer with such dimensions will be 0.26 μm^2, while the silicon island waveguide mode cross-section will be around 0.16 to 0.2 μm^2. Hence the coupling losses at the first MPSi layer interface will not exceed 30%. It is expected that by reducing the porosity of the MPSi layer (i.e., by reducing the macropore diameters) near both interfaces of the MPSi layer it is possible to increase the potential transmittance of such a structure over 75%. The same as for simple symmetries of MPSi arrays discussed previously, the macropores in the MPSi arrays of figure 6.13 can be filled by metal. In this case the cross-coupling between neighbor silicon island waveguides is expected to be even smaller (i.e., the ratio of waveguide mode area to unit cell area can be increased). For such an array the potential transmittance is expected to exceed 85%, which is suitable for most of applications of IR narrow-band-pass, band-pass, or band-blocking filters.

It should be noted that in all discussions in this section the coupling efficiency at the first interface of the MPSi layer was calculated as a ratio of the waveguide mode area to the MPSi array unit cell area. However, as it follows from expression (6.7) or (6.7a) the reflectance at the air/silicon boundary (as well as the reflectance at the silicon/air boundary during outcoupling at the second MPSi layer interface) strongly degrades the coupling (and outcoupling) efficiency. Due to the high refractive index of silicon, reflection losses are expected to reach 31% at each MPSi layer interface at normal incidence. It means that even for optimized MPSi array symmetries (like the ones shown in figure 6.14) the transmittance through such an MPSi array would not exceed 30 to 40%. Fortunately, the high reflection loss problem can be solved by coating of both surfaces of the MPSi layer with an antireflection coating. It can be a single-layer antireflection coating, like silicon monoxide (see section 3.1.1 of this book) or multilayer antireflection coating (see section 3.1.2 of this book).

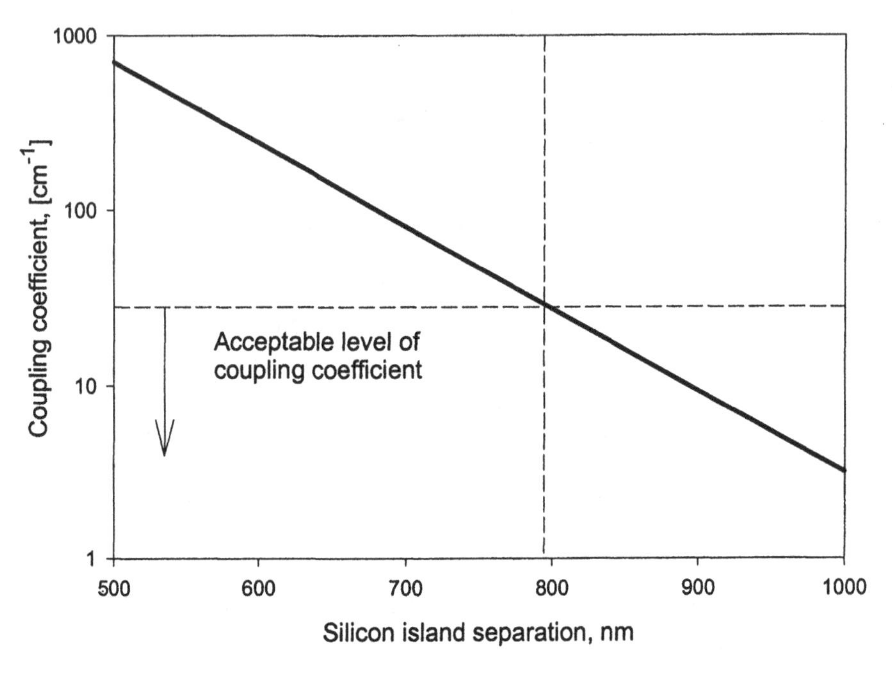

Figure 6.14. Numerically calculated dependences of cross-coupling coefficients of fundamental waveguide modes on the silicon island separation for the MPSi layer of figure 6.13a.

Special precautions should be taken in order to coat just the MPSi layer interfaces, not the macropores walls. Chemical vapor deposition seems to be prohibited as the method of the antireflection layer coating. From another point of view, magnetron sputtering should provide enough directivity. As was shown in sections 3.1.1 and 3.1.2, a correctly designed antireflection coating can suppress reflection losses almost completely so that the maximum potential transmittance of such an MPSi layer will be equal to the ratio of waveguide mode area to the MPSi array unit cell area.

Another issue that should be addressed during the design of narrow-band-pass, band-pass, or band-blocking filters is the value of effective reflective index modulation in silicon island waveguides. As was shown in Chapter 3 of this book, the bandwidth of the Bragg reflection peak is proportional to the amplitude of refractive index modulation in multilayer or waveguide (see figure 3.10). From another point of view there is a maximum obtainable variation in macropore diameters that can be tolerated while preserving an MPSi array ordering through its depth. This limitation will be discussed in more detail in section 6.4. Although precise limits of maximum obtainable variations in macropore diameters have not been proven to be found, we can assume maximum tolerable pore diameter variation to be around 30%. Maximum obtainable effective refractive index variations in this case can be approximated as follows:

$$\delta n^*_{\max} \approx \frac{\partial n^*}{\partial h}\delta h_{\max} \tag{6.9}$$

where h is the silicon island waveguide characteristic cross-section (for example, for circular waveguides h is the diameter of waveguide) and $\delta h_{\max}$ is the variation of this characteristic cross-section, corresponding to the 30% change in macropore diameter. While both dn^*/dh and $\delta h_{\max}$ strongly depend on MPSi array symmetry and structure, dn^*/dh also depends on wavelength. In first approximation we can consider these parts separately. The first part (dn^*/dh) strongly depends on how far the waveguide parameters are from the waveguide mode cut-off conditions.

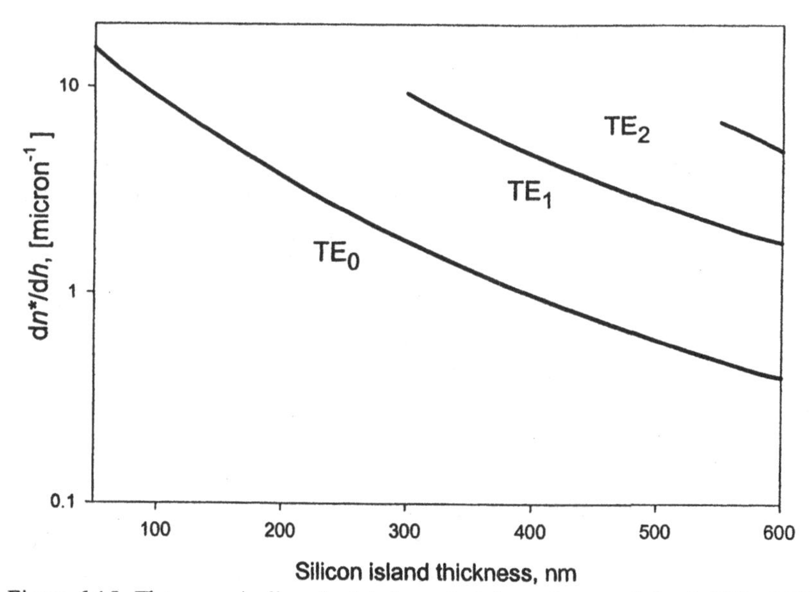

Figure 6.15. The numerically calculated spectral dependences of the dn^*/dh of TE-polarized waveguide modes versus silicon island thickness for the MPSi array of figure 6.4a.

As an example, in figure 6.15 the numerically calculated spectral dependences of the dn^*/dh of TE-polarized waveguide modes versus silicon island thickness are given for the MPSi array of figure 6.4a having silicon dioxide layer thickness of 275 nm. The wavelength was assumed to be 1550 nm. The numerically calculated dependences of effective refractive indices of waveguide modes for the same MPSi structure were given in figure 6.5. One can note that the absolute value of dn^*/dh decreases by almost two orders of magnitude with the increase of the silicon island cross-section from 50 to 600 nm. As was shown above, the optimum (from the viewpoint of coupling losses and cross-coupling suppression) value of the silicon island cross-section for such an MPSi structure is around 275 nm. According to figure 6.15 it corresponds to the value of dn^*/dh about 2.5 μm^{-1}.

The variation of this characteristic silicon island cross-section corresponding to the 30% change in macropore diameter, δh_{max}, is unique for each MPSi array. To get the filling for it, let us estimate its value for a nickel-filled cubic MPSi array with period (pore-to-pore distance of 1500 nm), near-square pores (after the oxide layer growth step) of 750x750 nm and a 300 nm silicon dioxide layer covering the pore walls (the same as for figure 6.12). Simple geometrical calculations give us a value of δh_{max} of about 200 nm that is, the maximum obtainable effective refractive index variations according to (6.9) will be around 0.5. According to figure 3.10, for such an MPSi array, the maximum bandwidth of Bragg reflection peak will be about 150 to 200 nm.

The situation is different for advanced symmetry MPSi arrays like the one shown in figure 6.13. In this case the macropore sizes are considerably smaller than the silicon island cross-section (unlike the MPSi array of figure 6.4, where macropores are bigger than the silicon island). For example, for the structure of figure 6.13a, optimized for 1550 nm wavelength, it was shown that macropore diameters after oxidation should be around 180 nm. The 30% modulation of such pore diameters will correspond to about a 20% modulation of the silicon island cross-section. In this case the maximum obtainable effective refractive index variations according to (6.9) will be about 0.2. According to figure 3.10, we can obtain that for such an MPSi array maximum bandwidth of Bragg reflection peak will be about 50 to 70 nm. It means that MPSi arrays with advance symmetry offer considerably higher potential transmission (due to lowering of coupling losses) but are disadvantageous from the viewpoint of the maximum obtainable Bragg reflection peak bandwidth.

To illustrate the expected advantages of the spectral filter based on this design with respect to common interference filters, numerical calculations are presented in figures 6.16 and 6.17. In figure 6.16a, the numerically calculated transmittance spectra through a seven-cavity narrow-band-pass multilayer dielectric filter (having a structure similar to the narrow-band-pass filter of figure 3.26) are presented for normally incident, 10- and 15-deg. tilted plane-parallel beams. The wavelength shift of the pass-band position together with the degradation of the pass-band shape, are shown. In figure 6.16b, the normalized transmittance spectra through a multiple-cavity MPSi-array-based filter (having a structure similar to the one in figure 6.4a) are presented for normal incident, 20° and 30° tilted plane-parallel beams. Transmittance is presented in normalized form, since the maximum transmittance is defined by the particular MPSi layer structure that was discussed above. It follows from figures 6.16a and 6.16b that the narrow-band-pass filter based on an MPSi array with coherently modulated macropore diameters will give the opportunity to use multiple cavity narrow-band-pass filters at different angles of incidence (±40 deg. at least), contrary to the interference-based filters discussed in section 3.6.3 of this book.

In figure 6.17a, the numerically calculated transmittance spectra through the narrow-band-pass multilayer dielectric filter of figure 6.16a are presented for normally incident beams with different divergences: plane-parallel beam (0 divergence angle) and Gaussian beams with 10 and 20 deg. divergence angles. The degradation of both the pass-band shape and the out-of-band rejection that are common to multiple-cavity all-dielectric multilayer FP filters are demonstrated. In figure 6.17b, the calculations for normalized transmittance spectra through the multiple-cavity MPSi filter of figure 6.16b are presented for 0, 20, and 40 deg. divergent, normally incident Gaussian beams. It follows from figures 6.17a and 1.17b that the narrow-band-pass filter based on an MPSi array with coherently modulated macropore diameters will provide similar advantages over multilayer filters at convergent or divergent beams up to about ±40 deg.

Similar advantages in performance are expected for band-pass and band-blocking filters based on MPSi arrays, discussed previously, since the transmittance shape in such filters is governed by the same interference phenomenon in the quarter-wave stack. Based on the above design considerations, MPSi-based filters can be designed to operate at anywhere from the 1.2 to 6 micron wavelength range since silicon is transparent at such wavelengths (see figure 6.2) and can serve as an waveguiding layer.

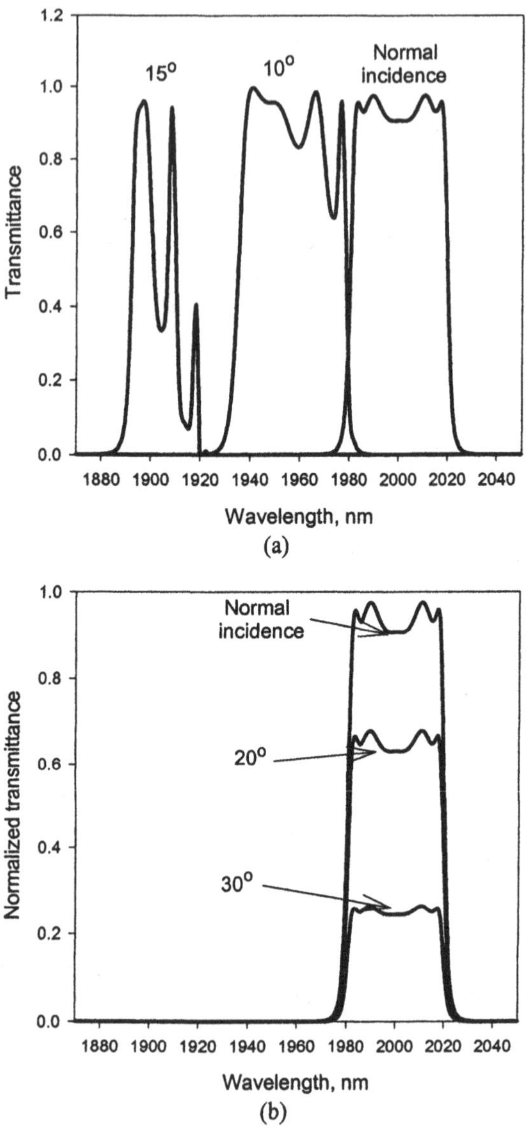

Figure 6.16. Numerically calculated transmission spectra through a quarter-wave multilayer stack (a) and through an MPSi filter (b) for different angles (light incident at 0, 10 and 15 deg. form normal) of incidence.

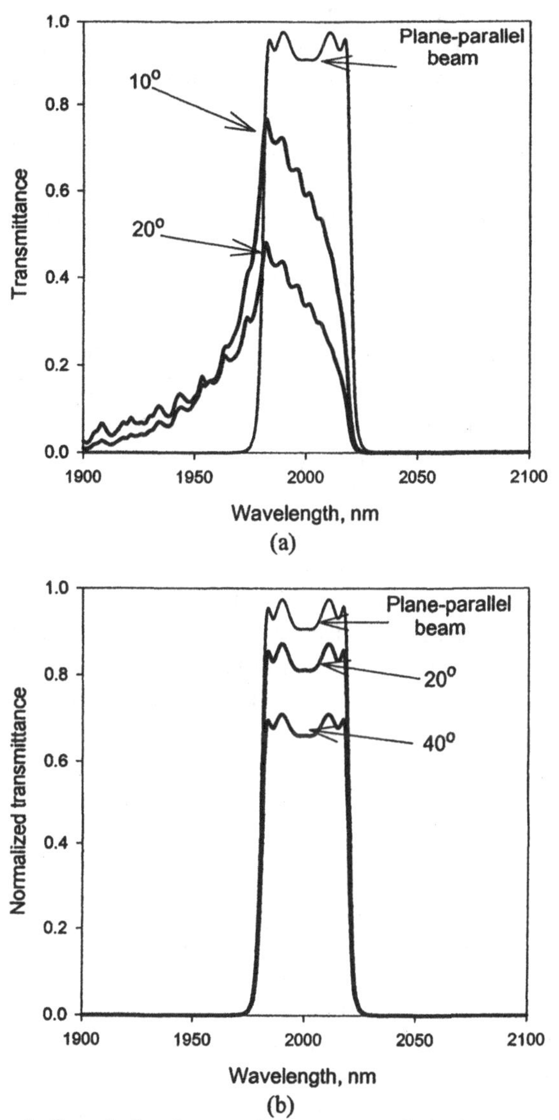

Figure 6.17. Numerically calculated transmission spectra through a quarter-wave multilayer stack (a) and through an MPSi filter (b) for different divergences of Gaussian beams.

6.3 MPSi-Based Materials for Filters Operating in the Visible Spectral Range

The spectral filters based on the MPSi layer structures discussed above have significant advantages over dielectric multilayer-based narrow-band-pass, band-pass, or band-blocking filters. However, these designs cannot be directly transferred into the visible and near IR spectral ranges (400 to 1100 nm wavelengths) due to the absorption of the silicon (see figure 5.4), which will cause unreasonable propagation losses in silicon island waveguides. However, visible and near IR spectral ranges are of great research commercial importance and omnidirectional narrow-band-pass, band-pass, and band-blocking spectral filters are clearly needed for these wavelengths.

An ordered array of waveguides with a coherently modulated cross-section made of materials that are transparent at visible spectral range (similar to the MPSi layer discussed above) would serve such purposes. Silicon, when thermally oxidized, from another point of view, forms silicon dioxide, which is transparent down to deep UV and far UV ranges. It is possible to oxidize the MPSi layer completely through all its depth, due to the porous structure of the MPSi layer. A *completely oxidized MPSi layer* (*COMPSi*) can be considered as an array of silicon dioxide waveguides (similar to an untreated MPSi layer at IR wavelengths, discussed in the previous section). The above theory can be directly transferred to completely oxidized MPSi layers, except for some normalization caused by a lower refractive index of silicon dioxide than that of silicon. In particular, the reflection losses during coupling should be about 4% at each interface of a freestanding COMPSi layer that is, coefficient 0.69 in expression (6.8) should be replaced by 0.96. The acceptance angle of silicon-dioxide-based waveguides should be also considerably less than that of silicon. However, to define it we need to consider other factors.

As with an untreated (unmodified) MPSi layer, silicon dioxide waveguides in a COMPSi layer cannot be considered independent of each other. In order to obtain visible-range, omnidirectional, narrow-band-pass, band-pass, or band-blocking filters based on the COMPSi layer, the structure of the COMPSi layer should be modified to suppress cross-coupling between neighboring silicon dioxide waveguides. This can be done in several ways, which are discussed below; however, none of them is as simple as the method used for IR filters.

The first method for modifying the COMPSi layer structure is based on nonuniform refractive index distribution in thermally grown silicon dioxide along its depth. Thermally grown silicon dioxide in general is a mixture of SiO_x with a different x: 0,1,2 (silicon, silicon monoxide, and silicon dioxide). Pure silicon dioxide, SiO_2, has the lowest refractive index of ~1.46 at visible and near IR spectral ranges and has the lowest absorption at these wavelengths. The presence of silicon (the refractive index of which is plotted in figure 5.4) and silicon monoxide (n = 1.7) in thermally grown silicon dioxide leads to the increase of both the refractive index of thermal oxide and the optical absorption of thermal silicon dioxide with respect to pure silicon dioxide at 400 to 1100 nm wavelengths.

From another point of view, the concentrations of silicon and silicon monoxide in thermally grown silicon dioxide are minimal at the surface and increase with the depth. The silicon and silicon monoxide concentration profiles depend on conditions of thermal oxidation like the temperature of the oxidation furnace and the concentration of oxygen and other gases in the oxidation furnace during the thermal oxidation process. In general, silicon and silicon monoxide impurities are minimal in thermal silicon dioxide obtained at high temperatures (1200 to 1300°C) in wet oxygen atmosphere, while maximal in thermal silicon dioxide, grown at lower temperatures (900 to 1000°C) in dry oxygen atmosphere. Hence, after complete thermal oxidation the MPSi layer under specific conditions, the silicon, and silicon monoxide impurities gradient will be created across the silicon dioxide islands in the COMPSi layer. As an illustration, in figure 6.18 the plot of expected refractive index distribution across the silicon dioxide island cross-section is presented for the COMPSi layer.

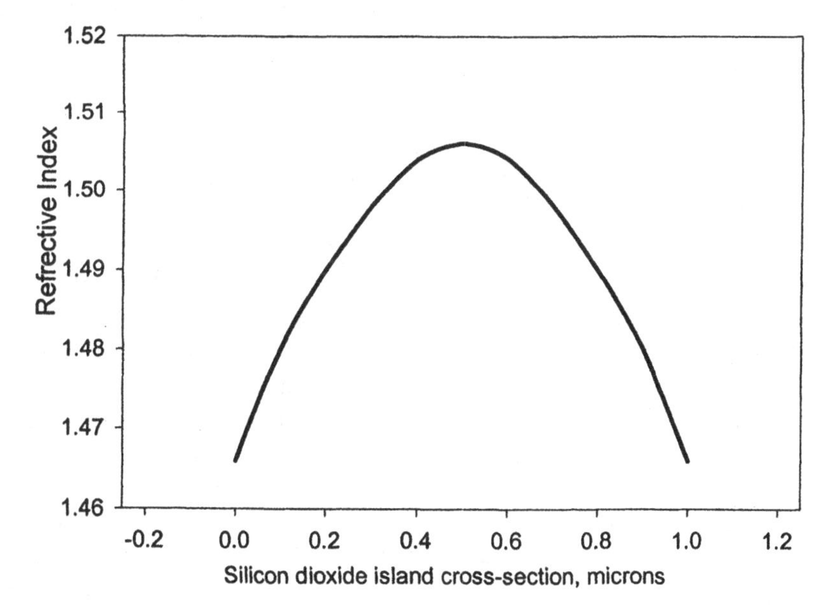

Figure 6.18. Refractive index distribution across the silicon dioxide island cross-section in the COMPSi layer.

Figure 6.19. Schematic view of the COMPSi layer cross-section for the array of an uniform cubic lattice with near-square pores in initial MPSi layer.

The two-dimensional distribution of the refractive index in such a COMPSi layer is schematically shown in figure 6.19. The pores are filled with black; the COMPSi host is filled with a dark color where the concentration of silicon and silicon monoxide impurities is higher (and through that the refractive index is higher) and with a light color where the concentration of silicon and silicon monoxide impurities is lower (and through that the refractive index is lower). The initial MPSi layer (before thermal oxidation) was assumed to have the cubic symmetry of figure 6.4a and near-square pores.

It is known that structures with such an index gradient can support a waveguide mode – the so-called gradient-index waveguides discussed in section 2.6.2 of this book. The light in such waveguides will be confined in the silicon dioxide islands better than in uniformly oxidized MPSi layer; hence the cross-coupling between neighboring silicon island waveguides should be suppressed at some extent. If the original MPSi layer has been made in the form of an ordered macropore array with modulated pore diameters, the COMPSi layer obtained by thermal oxidation of the MPSi layer should exhibit similar advantages for spectral filtering applications as the MPSi layer in the IR wavelength range, discussed in previous section. Although silicon monoxide and silicon impurities, beyond increasing the refractive index (essential for such an COMPSi layer design), will also caused increase in silicon dioxide absorption and hence waveguide propagation losses, this absorption is expected to play a minor role since (see the discussion in the previous sections of this book) the thickness of

COMPSi layer (hence the length of each silicon dioxide waveguide) will not exceed 300 μm. It is expected that the COMPSi layer will be suitable as a base for narrow-band-pass, band-pass, or band-blocking filters through the 400 to 2500 nm wavelength range.

Let us estimate the expected performance of the COMPSi layer. From figure 6.18 we can conclude that the acceptance angle silicon dioxide waveguides will be considerably less than that of silicon island waveguides in the MPSi layer due to refractive index contrast between a core and a cladding that is more than an order less. As follows from expression (2.60) and figure 6.18, the acceptance angle of the silicon island waveguide will be within the 45° range. Hence, the term *omnidirectional* applying to the COMPSi layer has less meaning as for the MPSi filter in the IR range. However, such acceptance angles still exceed by far the angles that can be tolerated by conventional multiple-cavity interference filters (see figures 3.27, 6.16, and 6.17). Another issue of concern for COMPSi filters is the maximum obtainable transmittance (or the coupling losses) at the first COMPSi interface. To estimate it we need to get first the approximate COMPSi layer structure.

As with the MPSi layer discussed in the previous section, the overall transmittance of the COMPSi layer will be determined by the particular macropore array geometry (i.e., by the relation between the wavelength of light, pore diameter, pore diameter modulation, pore-to-pore distance, etc.), macropore array symmetry, refractive index of pore-filling material (if any), and the refractive index distribution inside the COMPSi layer. The same arguments that were given during the MPSi layer discussion can be applied to the COMPSi layer: the advanced symmetries of the COMPSi layer (similar to the MPSi layer symmetries of figure 6.13) are expected to provide minimal coupling losses at the first COMPSi layer interface. COMPSi layers of such symmetries can be obtained by complete thermal oxidation of MPSi layers of figure 6.13. The same as for IR silicon island waveguides, silicon dioxide waveguides in COMPSi layer should be preferably single-mode. This requirement defines the particular sizes of the macropores.

In figure 6.20 the numerically calculated spectral dependences of TE modes effective refractive indices are given for COMPSi layer have cubic macropore symmetry with near-square pores cross-section. The macropore array period (which is equal to pore-to-pore distance for cubic arrays) was assumed to be 2200 nm, while the pores were assumed to be 910x910 nm. As follows from figure 6.20, for the COMPSi layer of given sizes, silicon dioxide waveguides became single-mode at the wavelength of approximately 620 nm. Next, we need to estimate waveguide mode cross-section. The numerically calculated waveguide-mode electric field distribution cross silicon dioxide island cross-section at the wavelength of 700 nm is given in figure 6.21.

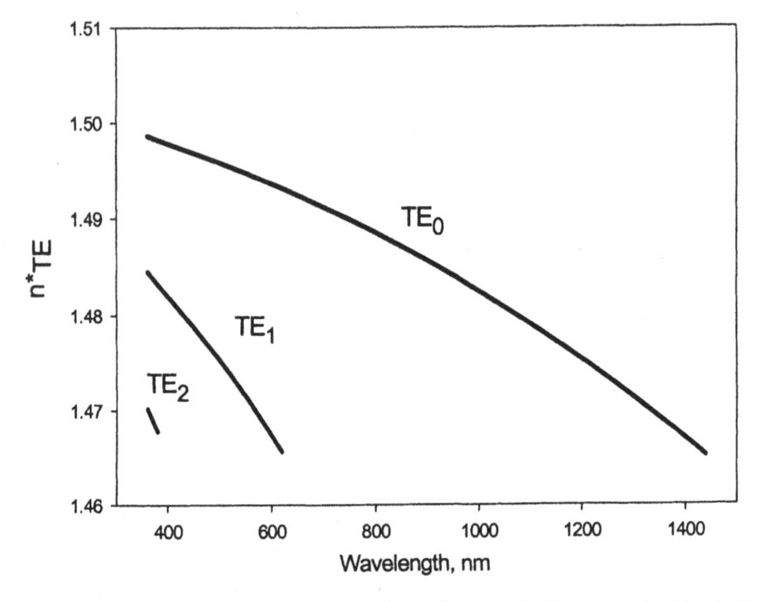

Figure 6.20. Numerically calculated spectral dependences of effective refractive indices of TE polarization waveguide modes for the COMPSi layer having 2200 nm pore-to-pore distance, 910x910 nm square pores.

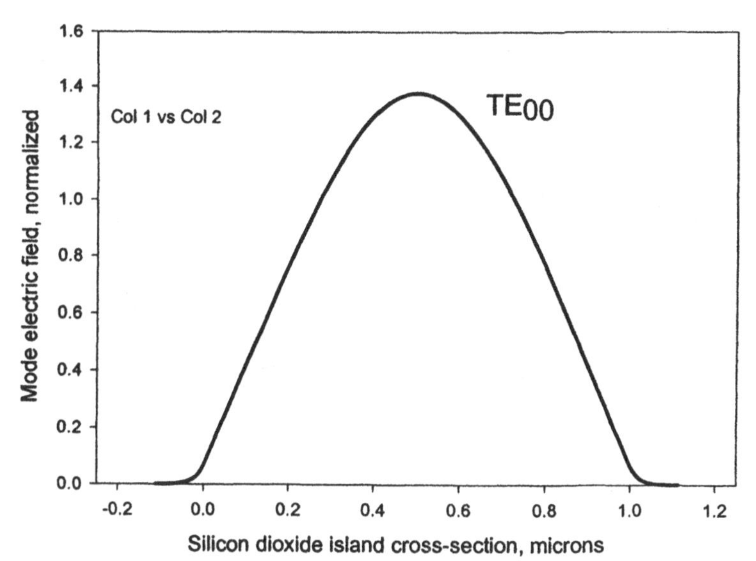

Figure 6.21. Numerically calculated waveguide mode electric field distribution across silicon dioxide island cross-section.

As follows from figure 6.21, the waveguide mode area in such a gradient index waveguide is about 30 to 40% less than the area of silicon dioxide island. Hence, for the given sizes of COMPSi layer the coupling efficiency at the first COMPSi interface is expected to be about 20%. As with the MPSi layer, the porosity can be optimized near the first and second interfaces of COMPSi layer to maximize the coupling and outcoupling efficiencies, so this value is expected to be at the 25 to 30% range. However, we need to check the cross-coupling between neighbor silicon dioxide waveguides.

In figure 6.22 the numerically calculated dependences of cross-coupling coefficients of fundamental waveguide modes on the silicon dioxide islands separation are presented for the COMPSi layers of cubic symmetry with near-square pores (1), of advanced hexagonal symmetry (like in figure 6.13a) (2), and of advanced hexagonal symmetry with pores filled with nickel (3). The calculations were done for 700 nm wavelength. The silicon dioxide island cross-section was assumed to be 1090 nm according to the single-mode requirement for this wavelength (see figure 6.21).

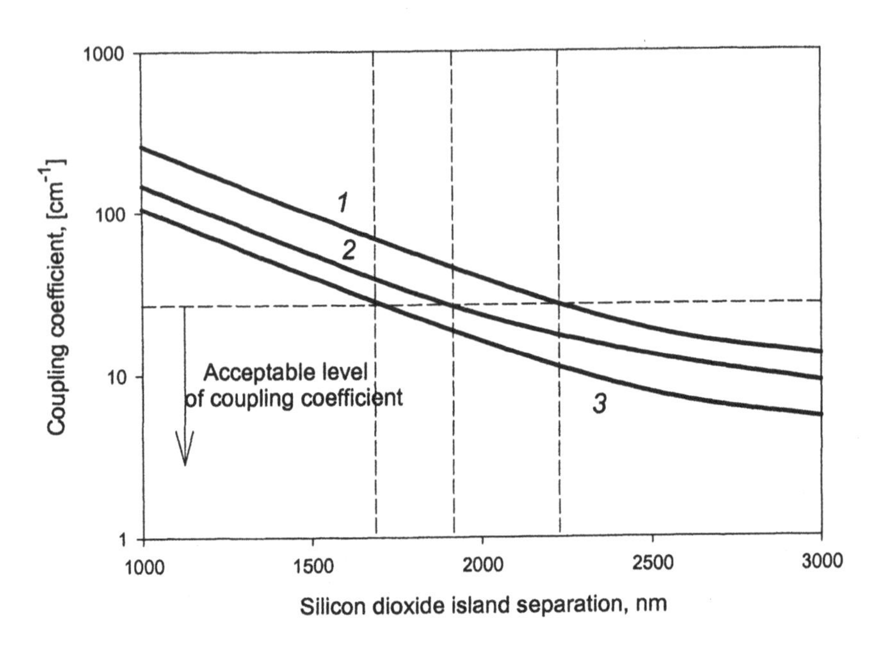

Figure 6.22. Numerically calculated dependences of cross-coupling coefficients of fundamental waveguide modes on the silicon dioxide island separation for the COMPSi layers of cubic symmetry with near-square pores (1), of advanced hexagonal symmetry (like in figure 6.13a) (2), and of advanced hexagonal symmetry with pores filled with nickel (3).

The acceptable level of coupling coefficient for the COMPSi layer is the same as for the MPSi layer (20 cm^{-1}), since the thickness of the COMPSi layer is expected to be within the same 50 to 300 μm range. As follows from figure 6.22, the cross-coupling coupling coefficient reaches an acceptable level for the cubic symmetry near-square pore COMPSi array at approximately 2500 nm silicon dioxide island separation; for the advanced hexagonal symmetry circular pore COMPSi array at approximately 1900 nm; and for the advanced hexagonal symmetry circular filled-with-nickel pore COMPSi array at approximately 1700 nm. These values correspond to the maximum potential transmittance through COMPSi arrays at 700 nm wavelength in the range of 15 to 17% for cubic symmetry near-square air-filled pores, 30 to 35% for advanced hexagonal symmetry circular air-filled pores, and 35 to 40% for advanced hexagonal symmetry circular filled-with-nickel pores. These values are lower than that of the MPSi arrays of the same symmetry because of worse localization of waveguide modes in silicon dioxide island cores, caused by a considerably smaller refractive index contrast. These values can be enlarged by adiabatic decreasing of pore diameters near both interfaces of the COMPSi layer to 20 to 25% for cubic symmetry near-square air-filled pores, 40 to 50% for advanced hexagonal symmetry circular air-filled pores, and 50 to 60% for advanced hexagonal symmetry circular filled-with-nickel pores. It means that in order for such a COMPSi layer to serve as the base for narrow-band-pass, band-pass, or band-blocking filters, metal filling of the pores is necessary.

As with the MPSi filters, the reflection losses occur at both COMPSi layer interfaces. As was mentioned above, these losses are much smaller and can be estimated as 4% per each interface. However, since the coupling losses are expected to be higher in the COMPSi layer, it can be still necessary to deposit antireflection coatings on both COMPSi interfaces. It can be single-layer low refractive index coating (cryolite of magnesium fluoride) or multilayer antireflection coating, similar to the one in figure 3.3.

As was mentioned above, metal filling of the pores is essential to maximize transmittance through the COMPSi layer. Let us estimate the expected level of metal-caused propagation losses of silicon dioxide island waveguide modes for the cubic symmetry COMPSi array in figure 6.20. For different symmetries of COMPSi array we can expect similar dependences. In figure 6.23 the spectral dependences of optical losses in silicon dioxide island waveguides are presented nickel-filled pores. As with the MPSi layer, the losses for each waveguide mode are increasing with the increase of the wavelength due to the decrease of localization of waveguide modes in the centers of silicon dioxide islands, causing less interaction of the modes with pore-filling metal. As follows from figure 6.23, the losses reach an acceptable level (which is the same as for MPSi layers 10 cm^{-1}) near the second-order mode cut-off. However, unlike the MPSi case, the differences between losses of fundamental and first-order modes are considerably smaller. Still, the differences in loss coefficients between fundamental and higher-order modes are sufficient for the COMPSi layer to be used at the wavelengths, where silicon island waveguides become multimode.

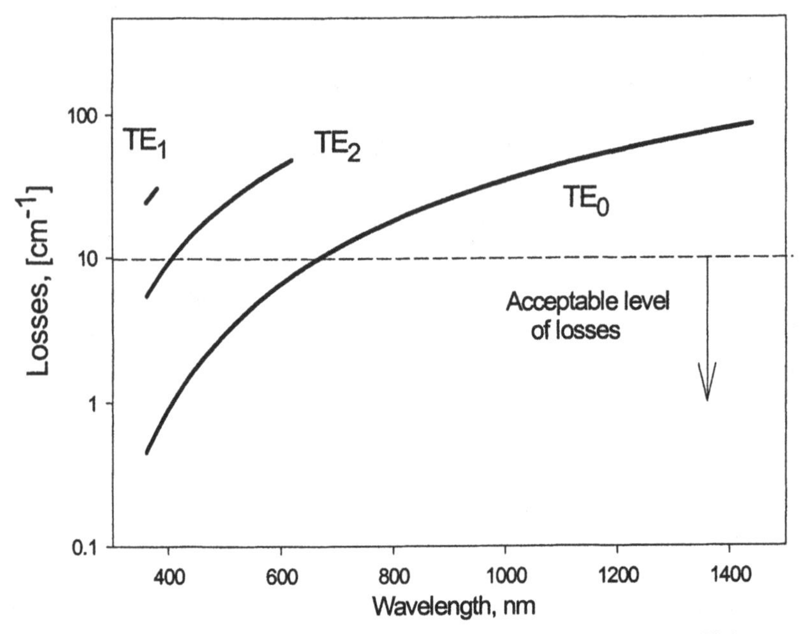

Figure 6.23. Numerically calculated spectral dependences of loss coefficients of TE polarization waveguide modes for the COMPSi layer of cubic symmetry with 2200 nm pore-to-pore distance and 910x910 nm near-square pores completely filled with nickel.

As with the MPSi layer case, the maximum achievable value of the effective reflective index modulation in silicon dioxide island waveguides is the important parameter of the COMPSi layer. Since the COMPSi layers are obtained from the MPSi layer by thermal oxidation, the maximum obtainable variations in the macropore diameters is around 30%. Equation (6.9) is general for any types of waveguides; hence it can be used equally for both MPSi and COMPSi layers.

As was mentioned above for the MPSi layers, (dn^*/dh) part of (6.9) strongly depends on how far the waveguide parameters are from the waveguide mode cut-off conditions. As an example, in figure 6.24 the numerically calculated spectral dependences of the dn^*/dh of TE-polarized waveguide modes are given for the nickel-filled COMPSi layer of figure 6.20. One can note that the absolute value of dn^*/dh decreases with the decrease of wavelength. As was shown above, the optimum (from viewpoint of coupling losses and cross-coupling suppression) wavelength range for such a COMPSi structure is around 600 to 700 nm. According to figure 6.24, it corresponds to the values of dn^*/dh about 0.007 μm^{-1}. These values are by more than order of magnitude less than that of the MPSi layer due to less refractive index contrast between the core and the cladding. However, the sizes of the silicon dioxide island waveguides are considerably larger than the sizes of the silicon island waveguide and the maximum variation of the characteristic silicon dioxide island cross-section δh_{max} in COMPSi arrays is also larger. As with the MPSi case, discussed in the previous section, δh_{max} is unique for each COMPSi array symmetry and structure. For example, for nickel-filled cubic COMPSi array with period (pore-to-pore distance of 2200 nm), near-square pores of 910x910 nm simple geometrical calculations give us the value of δh_{max} about 400 nm: the maximum obtainable effective refractive index variations according to (6.9) will be around 0.002. According to figure 3.10, for such a COMPSi array the maximum bandwidth of Bragg reflection peak will be about 5 to 10 nm at 600 to 700 nm wavelength range. This is much lower than for the MPSi layer.

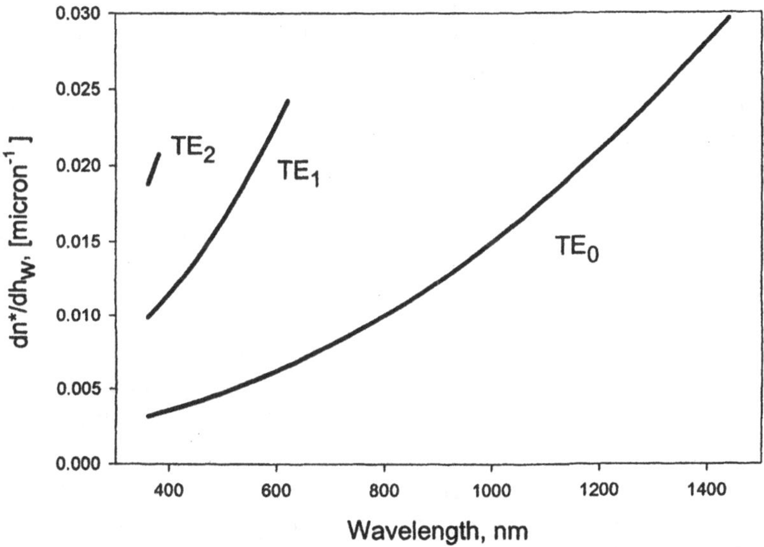

Figure 6.24. Numerically calculated spectral dependences of the derivations of the effective refractive indices of TE-polarized waveguide modes for the COMPSi layer of figure 6.20 with pores filled by nickel.

For advanced symmetry COMPSi arrays like that shown in figure 6.13 the maximum bandwidth of the Bragg reflection peak is even smaller due to the smaller pore diameters that are needed for such structure. For example, for the structure of figure 6.13a, it was shown in the previous section that 30% modulation of macropore diameters will correspond to about 20% modulation of silicon dioxide island cross-section. Hence, the maximum obtainable effective refractive index variations according to (6.9) will be about 0.001. According to figure 3.10, for such a COMPSi array the maximum bandwidth of Bragg reflection peak will be about 3 to 7 nm.

The lower value of the maximum achievable value of the effective reflective index modulation in silicon dioxide island waveguides of COMPSi layers not only causes lower bandwidth of the Bragg reflection peak but also requires the use more periods of pore diameter modulation to get the same level of reflectivity than MPSi layers (as an illustration, see figure 2.7). It can be that for COMPSi layers higher coupling losses or values of cross-coupling coefficients can be acceptable as a tradeoff to boost the value of δh_{max}.

Many other variations of MPSi layer structures can be imagined to modify its different performance aspects as an optical filter material. What has been presented here is the example of what could be done, not the finished development project. I believe that the material presented here is enough to give the reader an understanding of the new prospects that filters based on coherent arrays of independent waveguides open. Let us discuss now in more detail how such structures can be manufactured.

6.4 Fabrication of MPSi Layers with Coherently Modulated Pore Diameters

The base of the MPSi and COMPSi layers of sections 6.2 and 6.3 of this book is an ordered MPSi array with coherently modulated macropore diameters. The formation of MPSi arrays with constant pore diameters were discussed in the sections 5.4.1 (random macropore array) and 5.4.2 (ordered macropore array). There is no sense in repeating the materials of these sections here, so let us focus on the ways to get macropore diameters coherently modulated.

Modulation of macropore diameters can be done in many different ways. As was discussed in section 5.4.2 the average macropore diameters in the MPSi layer for a given combination of electrolyte solution and silicon wafer resistivity are proportional to applied current density during anodic etching of silicon. Hence, if applied current density is modulated in some fashion during the electrochemical etching process, the macropore diameters in the resultant MPSi layer will be also modulated. As an example, in figure 6.25 an SEM image shows the MPSi layer having modulated pore diameters. This MPSi layer was fabricated on p-doped silicon wafer with resistivity of 13

Ω cm, processed in electrolyte having composition {1 [HF]+2 [Alcohol]+8 [$HCON(CH_3)_2$])} at room temperature. In figure 6.26 the time dependence of applied current density during fabrication of the MPSi layer of figure 6.25 is given. One can clearly see that the macropore diameters indeed are modulated with MPSi layer depth. However, the step-like applied current density function (see figure 6.26) has not reproduced. Instead, the macropore diameters are smooth functions of MPSi depth.

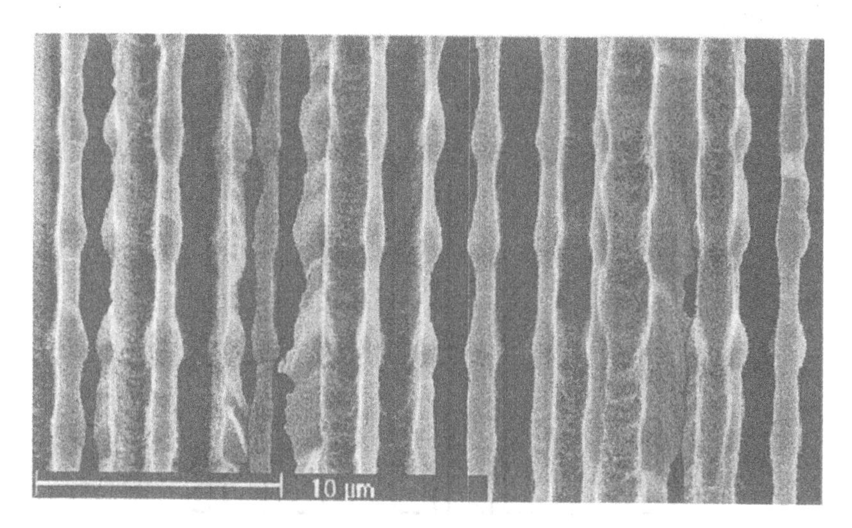

Figure 6.25. An SEM image of the MPSi layer with modulated with depth pore diameters.

In general, the steepness of the macropore diameter change is limited by the ~45° angle between the macropore wall and the direction of pore growth. This conclusion is based on experimental findings.

Several factors require special attention during the fabrication of an ordered MPSi array with coherently modulated macropores diameters. For example, the amplitude of applied current density modulation (and through that the macropore diameter modulation) should not exceed the critical value, above which the macropore ordering is not preserved. Under the modulations exceeding this critical value, the branching of macropores was observed at the interfaces between high- and low-porosity regions. This critical value is unique for any given combination of silicon resistivity, electrolyte composition, anodization process parameters, and symmetry and dimensions of MPSi array. It also depends strongly on the preferential pore-to-pore distance of silicon wafer, electrolyte, and process parameters (see the discussion in the section 5.4.2 of this book) from the MPSi array pitch. In the case of mismatch between preferential pore-to-pore distance and MPSi array pitch, loss of coherence in modulation was observed at some depth of the MPSi layer. This depth is a function of the mentioned mismatch (the less the value of mismatch, the deeper the coherently modulated MPSi array can be obtained). For the case of a perfect match of the MPSi array pitch to the preferential pore-to-pore distance, the critical value of pore diameter modulation is expected to be maximal for hexagonal symmetry of the MPSi array (see figure 6.4c) and minimal for advanced symmetries of the MPSi array (see figure 6.13).

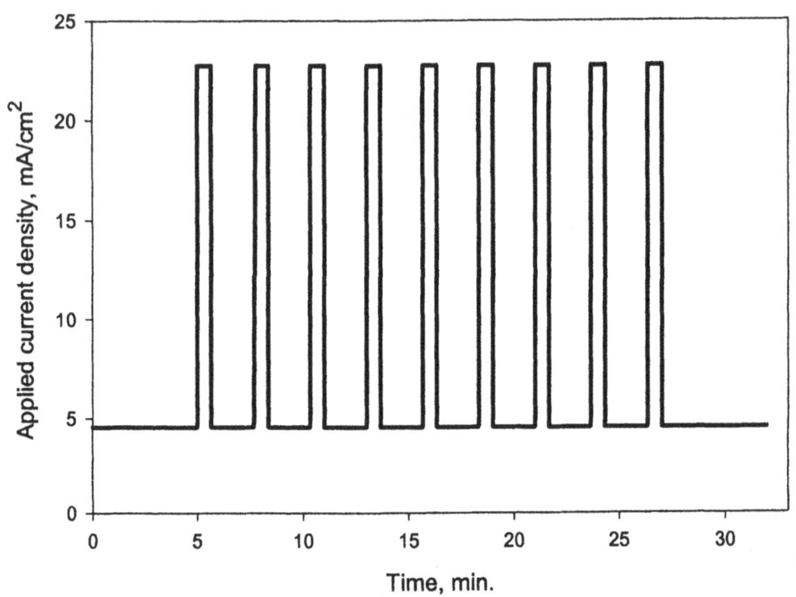

Figure 6.26. Time dependence of applied current density during fabrication of the MPSi layer of figure 6.25.

Another factor that should be addressed during fabrication of an ordered MPSi array with coherently modulated macropore diameters is the uniformity of the current density over the useful part of the silicon wafer under anodization. As an example of not optimized anodization conditions, figure 6.27 presents the experimentally measured spatial distribution of the MPSi growth rate across the silicon wafer opening in the anodizing chamber of figure 5.21. This MPSi layer was fabricated on a p-doped silicon with resistivity of 17 Ωcm, processed in electrolyte having the composition {1 [HF]+2 [Alcohol]+8 [$HCON(CH_3)_2$])} at room temperature and a current of 50 mA. The opening in the anodization chamber was 1.5 inches in diameter; the cathode was a 1x1 inch shape asymmetrically placed with respect to a wafer opening 2 inches apart. Such nonuniformity of the MPSi growth rate across the silicon wafer causes the different pore modulation periods at different spots of the MPSi wafer, causing the different wavelength positions of the Bragg peaks.

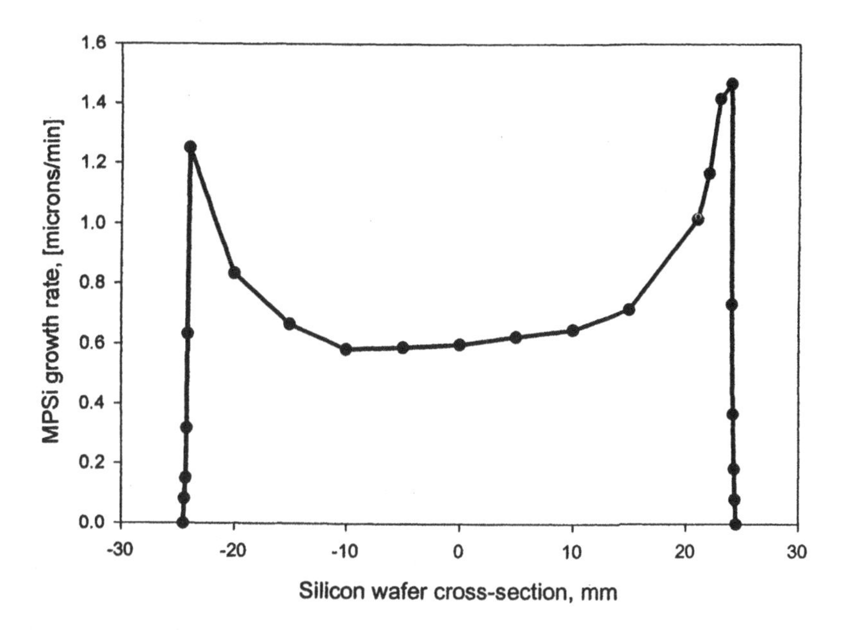

Figure 6.27. Spatial distribution of the MPSi growth rate across the silicon wafer opening in the anodizing chamber of figure 5.2.

To solve this problem the design of the anodization chamber should take into account the sizes and shapes of the cathode and silicon wafer opening, the spacing between the cathode and silicon wafer, the conductivity of electrolyte, and many other parameters. As an example, figure 6.28 presents the numerically calculated current density distribution inside the anodization chamber of figure 5.21 during the electrochemical etching of a silicon wafer. By proper process parameter design and optimization, the author succeeded in obtaining an MPSi growth rate uniform within 5% over all silicon wafer openings, and better than 2% over the silicon wafer opening except for a 5 mm area around the o-ring.

The modulation of current density during electro-chemical etching of silicon is the simplest way to coherently modulate macropore diameters. The macropore diameters depend on many other parameters like silicon doping type and density and illumination intensity (for n-type doped silicon wafers). Hence, by modulating these parameters during the silicon anodization process the MPSi array can be obtained with coherently modulated macropore diameters.

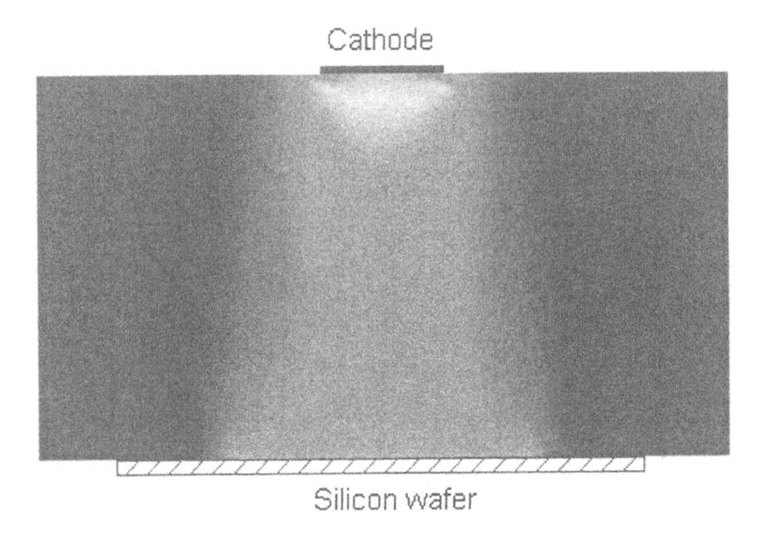

Figure 6.28. Numerically calculated current density distribution inside the anodization chamber of figure 5.21 during electrochemical etching of a silicon wafer.

Coherent modulation of macropore diameters with illumination intensity was demonstrated by [59]. Such a method is more complicated that the current modulation technique discussed above; however, it can provide controlled macropore diameter modulation distribution across the silicon wafer. Using this technique, interference filters with laterally varying wavelength of reflection (or transmission) can be constructed. This is virtually impossible for the current modulation technique. An approach that has been given for current-modulation in [60] gives considerably less control compared to the illumination modulation method.

Modulation of the doping density with the depth of the initial silicon wafer has not been demonstrated yet (to the author's knowledge) for macroporous silicon. However, such an approach has been successfully applied to the formation of refractive index super lattices in microporous silicon [61 and 62]. Due to the similarity of microporous and macroporous silicon formation this approach can be expected to be transferred for the fabrication of ordered MPSi arrays with coherently modulated macropores diameters. For microporous silicon this approach produced a very sharp interface between high- and low-porosity regions. If successfully applied to MPSi, this approach will guarantee the coherence of macropore modulation across anodized area of silicon wafer. The growth of a silicon wafer with modulated doping density across wafer depth is usually done by epitaxial growth of silicon layers on a substrate or by a CVD process.

6.5 Filling the Macropores with Metal

So far, the discussion has centered on the MPSi formation process. However, as was shown in sections 6.2 and 6.3, for an optimized optical performance of MPSi or COMPSi layers as a narrow-band-pass, band-pass, or band-blocking filters, the macropores should be filled with metal. It can be done in several ways.

The first method of filling of the pores with metal is an adaptation of standard electroplating techniques. The problem here is that in both MPSi and COMPSi layers the pore walls are covered by thermal silicon dioxide, which is insulating. From another point of view, electroplating requires electrically conductive surfaces to promote metal nucleation and growth [63]. This problem can be solved by catalytically seeding the surface with a Pd compound and then electroless plate a seed layer through the pores with an oscillating flow of the solution. The alternative method is to use flowing gas to deposit the seed layer by means of organometallic compounds. Various alternating flow methods with pulsed current source deposition can then be used to electroplate on the seed layer. While it is possible that CVD could fill the pores completely using techniques common in composite materials, it is more likely that the plating techniques used in the semiconductor industry to fill interlayer vias and capacitors will be more appropriate and less costly. As an example of the feasibility of such a technique, in figure 6.29 the photo of the MPSi layer with macropores completely filled by electroplated permalloy is presented. One can see that macropores are filled with metal all the way down to the bottom of the MPSi layer without voids. However, after the pores are filled with plated metal, the silicon (for MPSi) or silicon dioxide (for COMPSi) islands are usually coated by metal, which makes such layers completely opaque. A mechanical or chemical-mechanical polishing process can be used then to remove metal layers from the surfaces of the filters.

Another approach for promoting electroplating into pores is to anodically bond an unoxidized, low-resistivity Si wafer to the top of MPSi layer to act as a “handle wafer” and seed platform for bottom-up plating. After the plating is completed, the handle wafer can be etched away from the filter area in hot KOH. The remaining handle wafer will then form a "frame" on the filter that will add to the robustness.

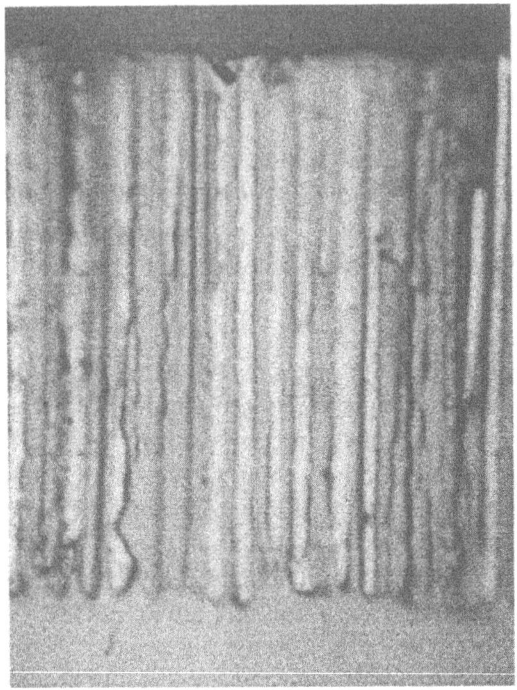

Figure 6.29. Photo of the MPSi layer with macropores completely filled by electroplated permalloy.

Another method of filling of macropores with metal has been demonstrated in [64]. In this paper the lead casting of a silicon mold was done. This method is considerably simpler than the electroplating technique discussed above. However, it has limitations. For example, the metal that can be used to fill the pores is limited by the requirement of a low melting point. In addition, during the die casting process stresses can developed in the MPSi layer, leading to cracking, although in [64] it was shown how it can be overcome.

6.6 Fabrication of MPSi -Based Spectral Filters

The MPSi- or COMPSi-based narrow-band-pass, band-pass, or band-blocking filter fabrication process is in many details similar to the MPSi-based short-pass filter fabrication process discussed in the section 5.5 of this book. However, some modifications should be made to the steps shown in section 5.5 so it is worthwhile to describe the MPSi- or COMPSi-based narrow-band-pass, band-pass, or band-blocking filter fabrication process separately:

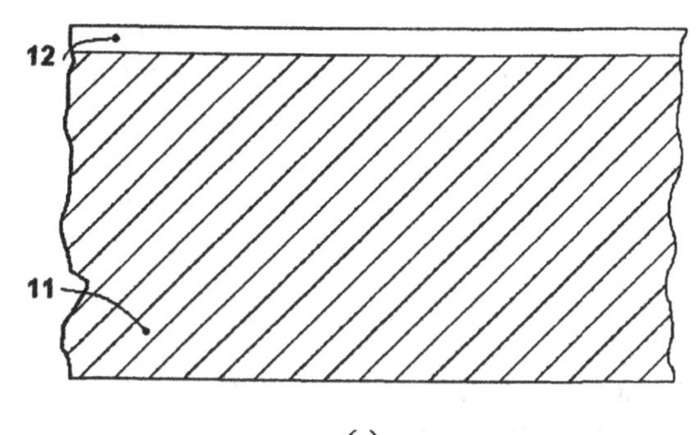

(a)

Figure 6.30a-j. A summary of the MPSi-based spectral filter fabrication process.

A single-crystal (100) orientation silicon wafer 11 (see figure 6.30a) chosen according to the above discussion of doping type and density should be covered with a thin film of chemically resistant material 1, for example, thermally grown or sputtered silicon dioxide or silicon nitride grown by LPCVD. Next, the etch-pit mask should be produced in the layer 12 (figure 6.30b) with the assistance of a conventional photolithography and subsequent chemical of RIE etching of layer 12 through the photoresist mask.

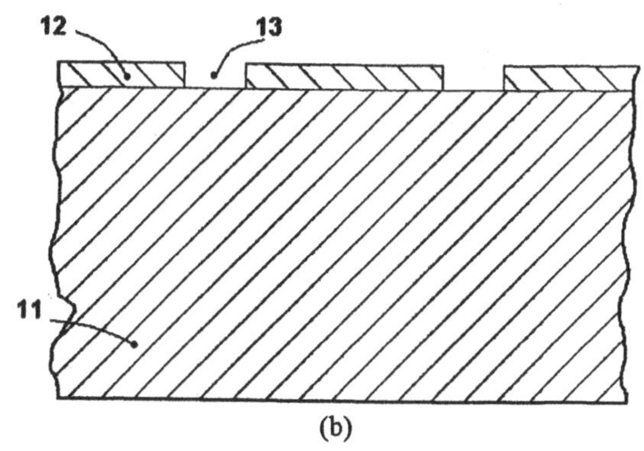

(b)

Next, the etch-pit mask should be transferred from layer 12 into the silicon wafer surface 11 (figure 6.30c) by chemical etching with the subsequent removal of the chemical resistant layer 12. Next, the first surface (with etch-pits) of the substrate 11 should be brought into contact with an electrolyte (figure 6.30d) having a composition carefully chosen according to silicon wafer characteristics and etch-pit array parameters (see the previous discussion). The substrate wafer 11 should be connected as an anode. The current density and illumination (if needed) should be applied between the substrate wafer 11 and the cathode placed in electrolyte. Through the anodization process, the macropores 15 will be grown starting from the tips of the etch-pits. The silicon anodization should be performed during the time required to form a macroporous layer with thickness predetermined by filter design considerations. During the anodization process the process parameters (like current density or back-side illumination intensity) should be modified in a periodic fashion, so the macropores will be formed with diameters that are modulated along their depth. After the anodization process is complete, the silicon wafer is removed from the electrolyte. The wafer should be carefully cleaned to ensure electrolyte is removed from the deep macropores. If organic addends were used in electrolyte, Piranha acid cleaning can be required.

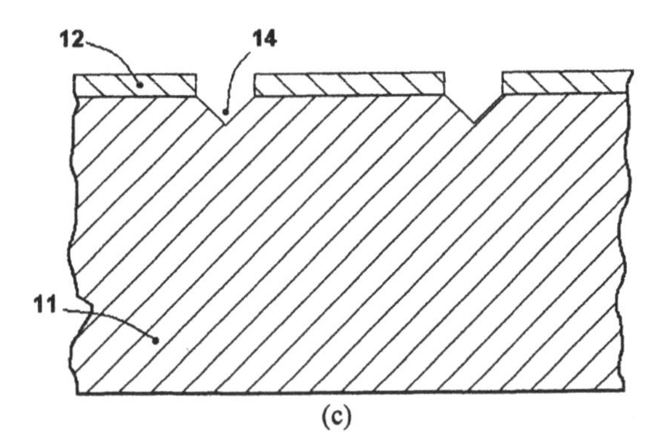

(c)

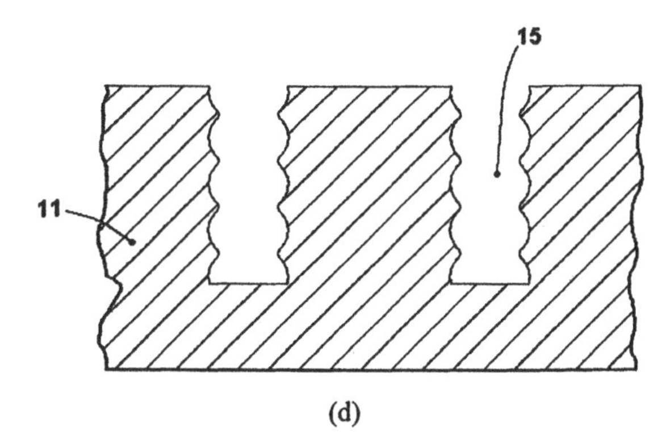

(d)

Next, the first surface of silicon wafer 11 is covered uniformly by one or more layers of transparent material (materials) (figure 6.30e). The first layer 18 can be grown by the thermal oxidation of said silicon wafer 11 in the oxygen atmosphere at the temperatures of 950 to 1300°C. The thickness of such a thermal silicon dioxide layer can be well controlled by the time of oxidation since this thickness at each temperature is proportional to oxidation time. In order to reduce oxidation-caused stress in the MPSi layer the wafer can be annealed in nitrogen atmosphere at the temperatures of 400 to 800°C. For COMPSi-based filters the macropore walls should be completely oxidized at the condition discussed previously in the text. All other layers (16, 17 of figure 5.38e) should be deposited by CVD or LPCVD. These layers (16, 17, 18) can be of silicon dioxide, silicon nitride, or any other material sufficiently transparent in visible and near IR spectral ranges.

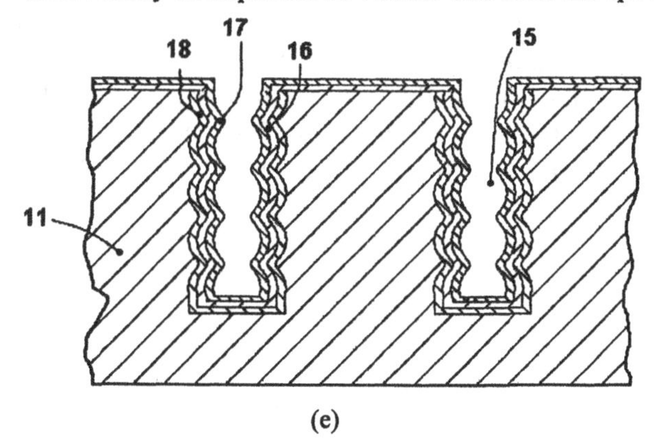

(e)

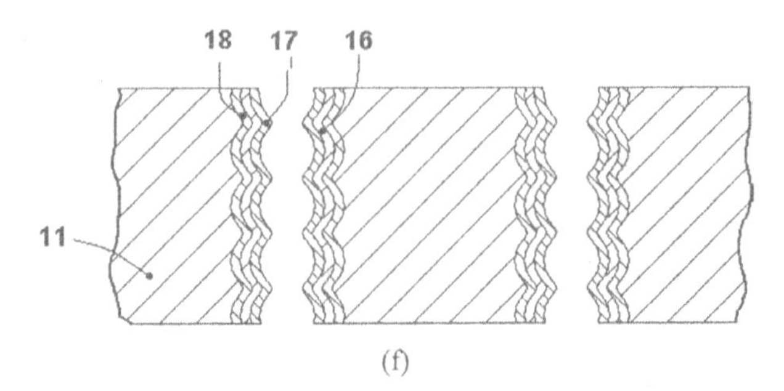

(f)

Next, the portion of silicon wafer 11 that does not have the MPSi layer is removed (figure 6.30f) through chemical etching or DRIE. The second surface of the MPSi layer then can be finished by mechanical polishing.

By following these fabrication steps, the functional spectral filter layer will be manufactured. However, as was noted in the previous discussion, metal filling of the pores can be necessary to optimize optical performance of the filter. In this case the filling of the pores with metal 19 (see figure 6.30g) can be accomplished by electroplating. The electroplating of the metal or alloy into pores can be done as follows. One of the surfaces of the freestanding MPSi layer can be placed into physical contact with the electrode of first polarity (it can be mechanically clamped onto the electrode surface, glued by conductive glue so that the distance between the surface of the filter wafer and the electrode surface does not exceed 10 to 20 microns). Then it can be placed into a plating bath filled with a plating solution having composition optimized so that the right metal or alloy is plated into the pores uniformly. The electrode of the second polarity is placed into the same plating bath at a predetermined distance from the MPSi filter wafer, and the voltage (or current) is applied between two electrodes. The current will flow through the pores only, since the MPSi (or COMPSi) wafer will be completely insulating.

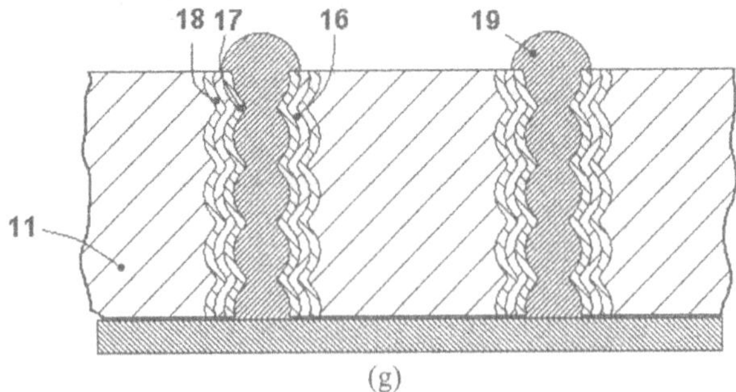

(g)

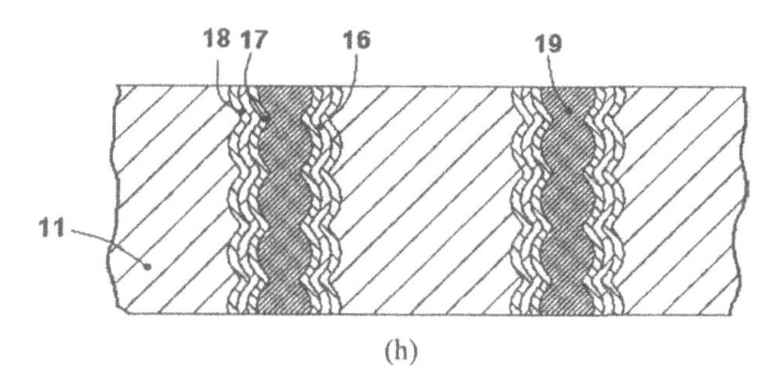

(h)

The metal after the pore-filling step most probably will cover also at least partially the areas of silicon or silicon dioxide islands at both surfaces of the wafer. For the spectral filter to be functional this excessive metal should be removed from both surfaces. The excessive metal can be removed from both surfaces of the spectral filter wafer by, for example, the mechanical or chemo-mechanical polishing of both surfaces of the MPSi (or COMPSi) wafer until the silicon (or silicon dioxide in the COMPSi case) host 11 is reached (see figure 6.30h).

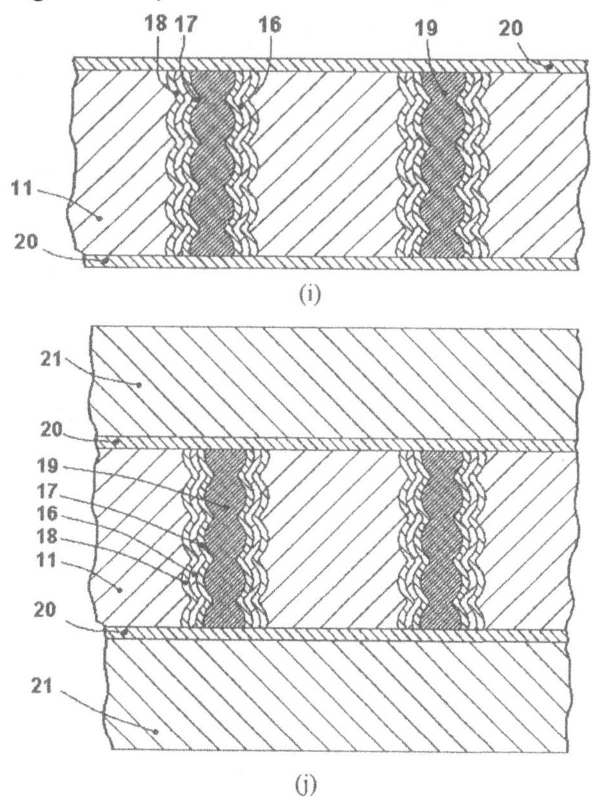

(i)

(j)

As it was mentioned in the previous discussion, an antireflection coating of silicon (or silicon dioxide) waveguide ends is strongly desirable to maximize the overall transmittance through the MPSi (or COMPSi) layer within its pass-band. In this case, the next step (see figure 6.30i) is to coat both surfaces of the MPSi (or COMPSi) layer by an antireflection layer or multilayer coating, which can be done through, for example, magnetron sputtering.

By following the above steps the functioning narrow-band-pass, band-pass, or band-blocking filter will be fabricated. However, it will not be robust mechanically. Reinforcing of such a layer can be required (figure 6.30g).

As an example of the MPSi structure, suitable for IR filters, figure 6.31 shows an SEM image of the MPSi layer with the structure like in Figure 6.4a. Figure 6.31a shows the back side of the MPSi layer, obtained after KOH etching of the unanodized part of silicon wafer in a coarse scale, while Figure 6.31 b shows an SEM image of the same side of the same wafer in a finer scale. Such a pore geometry was obtained in an DMS-based electrolyte. The orientation of the pore shape and the pore array was managed by the orientation of the etch pits on the silicon surface. The photolithographic mask was rotated by the 45 deg. with respect to the crystallographic axes of the silicon wafer.

As was mentioned at the beginning of this chapter, such filters are still at the research stage. There are still a lot of things that need to be proved and optimized. However, the author believes that the material presented in this chapter will be interesting to the reader and can give a fresh look into narrow-band-pass, band-pass, and band-blocking filter design and performance.

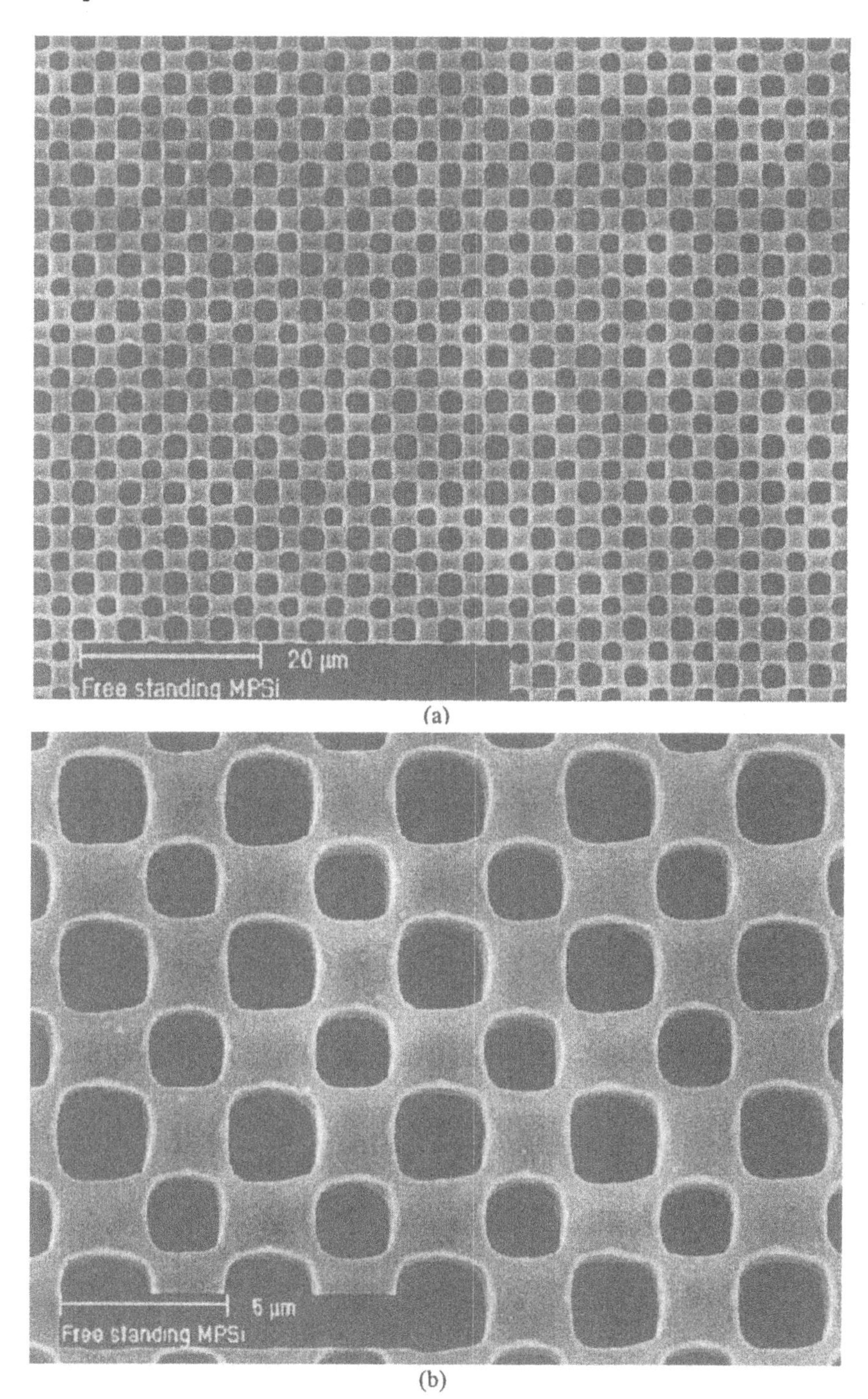

Figure 6.31. An SEM image of the MPSi layer with the structure like in Figure 6.4a.

References

[1]. Rouard P., *Contes Rendus de l'Academie de Science*, **195**, p. 896, 1932.

[2]. Bauer, G., *Ann. Phys.*, **19**, p. 434, 1934.

[3]. Strong J., *J. Opt. Soc. Am.*, **26**, p. 73, 1936.

[4]. Othons A., and Kalli K., *Fiber Bragg Gratings: Fundamentals and Applications in Telecommunications and sensing*, Artech House, 1999.

[5]. Okamoto K., *Fundamentals of Optical Waveguides*, Academic Press, 2000

[6]. Joannopoulos J.D., Meade R.D., Winn J.N., *Photonics Crystals: Molding the Flow of Light*, Princeton University Press, 1995.

[7]. McIntosh K.A., Mahoney L.J., Molvar K.M., McMahon O.B., Verghese S., Rothschild M., Brown E.R., "Three-dimensional metallodielectric photonic crystals exhibiting resonant infrared stop bands," *Appl. Phys. Lett.*, **70 (22)**, p. 2937, 1997.

[8]. Fink Y., Winn J.N., Fan S.H., Chen C.P., Michel J., Joannopoulos J.D., Thomas E.L., "A dielectric omnidirectional reflector," *Science*, **282 (5394)**, p. 1679, 1998.

[9]. Sharkawy A., Shi S.Y., Prather D.W., "Multichannel wavelength division multiplexing with photonic crystals," *Appl. Optics*, **40 (14)**, p. 2247, 2001.

[10]. Takano K., Nakagawa K., "Frequency analysis of wavelength demultiplexers and optical filters with finite 2-D photonic crystals," *IEICE Transactions on Electronics*, **E84C (5)**, p. 669, 2001.

[11]. Imada M., Noda S., Chutinan A., Mochizuki M., Tanaka T., "Channel drop filter using a single defect in a 2-D photonic crystal slab waveguide," *J. Lightwave Tech.*, **20 (5)**, p. 845, 2002.

[12]. Kosaka H., Kawashima T., Tomita A., Notomi M., Tamamura T., Sato T., Kawakami S., "Photonic crystals for micro lightwave circuits using wavelength-dependent angular beam steering," *Appl. Phys. Lett.*, **74 (10)**, p. 1370, 1999.

[13]. Avrutsky I., Kochergin V., Zhao Y., "Optical demultiplexing in a planar waveguide with colloidal crystal," *IEEE Photonics Tech. Lett.*, **12 (12)**, p. 1647, 2000.

[14]. Keilmann F., *Int. J. of Infrared and Millimeter Waves* **2**, p. 259, 1981.

[15]. Timusk T., Richards P.L., *Appl. Optics* **20**, p.1355, 1981.

[16]. Huggard P. G., Meyringer M., Schilz A., Goller K., Prettl W., "Far-infrared bandpass-filters from perforated metal screens," *Appl. Optics* **33**, p. 39, 1994.

[17]. Lehmann V., Stengl R., Reisinger H., Detemple R., Theiss W., "Optical shortpass filters based on macroporous silicon," *Appl. Phys. Lett.*, **78, 5**, p. 589, 2001.

[18]. Born M., Wolf E., *Principles of Optics*, 5th ed., Pergamon, 1975.

[19]. Yariv A., Yeh P., "*Optical Waves in Crystals: Propagation and Control of Laser Radiation,*" John Wiley & Sons, 1984.

[20]. Marcuse D., *Theory of Dielectric Optical Waveguides*, Academic Press, 1974.

[21]. Yeh P., *Optical Waves in Layered Media*, John Wiley & Sons, 1988.

[22]. Tervonen A., *Introduction to Glass Integrated Optics*, ed. S.I. Najafi, Artech House, 1992.

[23]. Najafi, S.I., *Appl. Opt.*, **27**, p. 3728, 1988.

[24]. Snitzer, E., "Cylindrical dielectric waveguide modes," *J. Opt. Soc. Amer.* **51**, p 491, 1961.

[25]. Bilodeau F., et al., *IEEE Photonics Tech. Lett.*, **7**, p.388, 1995.

[26]. Davis M.A., KerseyA.D., *Electronic Lett.*, **31**, p. 822, 1995.

[27]. Oulette F., *Optics Lett.*, **16**, p. 847, 1987.

[28]. Winic K.A., *Appl. Opt.*, **31**, p. 757, 1992.

[29]. Yamada M., Sakuda K., *Appl. Optics*, **26**, p. 3474, 1987.

[30]. Kogelnik H., Shank C.W., *Appl. Phys.*, **43**, p. 2327, 1972.

[31]. Yariv A., *IEEE Journal of Quantum Electronics*, **QE-9**, p. 919, 1973.

[32]. Sipe J.E., Poladian L., de Sterke C.M., *JOSA* A, **11**, p. 1307, 1994.

[33]. Cox J.T., Haas G., *J., Opt. Soc. Am.*, **48**, p. 677, 1958.

[34]. Macleod H.A., *Thin-Film Optical Filters*, 3rd ed., Institute of Physics Publishing, 2001.

[35]. Johnson P.B., Christy R.W., *Phys. Rev.* ***B***, **6**, **12**, 1972.

[36]. Yariv A., *Quantum Electronics*, 2nd ed., John Wiley & Sons, 1975.

[37]. Heavens O.S., Liddel H.M., *Appl. Opt.*, **5**, p. 373, 1966.

[38]. Turner A.F., Baumeister P.W., *Appl. Opt.*, **5**, p. 69, 1966.

[39]. Palik E.D., *Handbook of Optical Constants of Solids*, Academic Press, 1998.

[40]. Liddell H.M., *Computer-Aided Techniques for the Design of Multilayer Filters*, Adam Hilger, 1981.

[41]. Thelen A., *J. Opt. Soc. Am.*, **53**, p. 1266, 1963.

[42]. Turner A.F., *J. Phys. Radium*, **11**, p. 443, 1950.

[43]. J. N. Winn, et al., *Opt. Lett.* **23**, p. 1573, 1998.

[44]. Deopura M., Ullal C.K., Temelkuran B., Fink Y., *Optics Letters* **26 (15)**, p. 1197, 2001.

[45]. Kim S.H., Hwangbo C.K., *Applied Optics*, **41 (16)**, p. 3187, 2002.

[46]. Gallas B., Fisson S., Charron E., Brunet-Bruneau A., Vuye G., Rivory J., *Appl. Optics*, **40 (28)**, p. 5056, 2001.

[47]. Wang X., Hu X.H., Li Y.Z., Jia W.L., Xu C., Liu X.H., Zi J., *Appl. Phys. Lett.*, **80 (23)**, p. 4291, 2002.

[48]. US Patent 6,014,251 issued to A. Rosenberg et al., January 11, 2000.

[49]. Lehmann V., *J. Electrochem. Soc.*, **140 (10)**, p. 2836, 1993.

[50]. Pickering C., Beale M.I.J., Robbins D.J., Pearson P.J., Greef R., *J. Phys. C (Solid State Physics)*, **17, 35**, p. 6535, 1984.

[51]. Lehmann V., Foll H., *J.Electrochem. Soc.*, **137, 2**, p. 653, 1990.

[52]. Lehmann V., Gruning U., *Thin Solid Films*, **297**, p. 13, 1997.

[53]. Christopherson M. et al., *Mater. Sci. and Eng. B*, **69-70**, p. 194, 2000.

[54]. Propst E.K., Kohl P.A., *J. Electrochem. Soc.*, **14**, p. 1006, 1994.

[55]. Ponomarev E.A., Levy-Clement C., *Electrochem. and SoldiState Lett.*, **1**, p. 45, 1998.

[56]. Lehmann V., Ronnebeck S., *J. Electrochem. Soc.* **146 (8)**, p. 2986, 1999.

[57]. Allongue P. et al, *Appl. Phys. Lett.*, **67, 7**, p. 941,1995.

[58]. Ohji H. et al, *Sensors and Actuators*, **85**, p. 390, 2000.

[59]. Schilling J., Muller F., Matthias S., Wehrspohn R.B., Gosele U., Busch K., "Three-dimensional photonic crystals based on macroporous silicon with modulated pore diameter," *Appl. Phys. Lett.*, **78, 9**, 2001.

[60]. Hunkel D. et al., *J. of Luminescence*, **80**, p. 133, 1998.

[61]. Berger M.G., Arens-Fischer R., Frohnhoff St., Dieker C., Winz K., Munder H., Luth H., Arntzen M., Theiss W., (ed. Collins R.W., Tsai C.C., Hirose M., Koch F., Brus L.), "Formation and properties of porous Si superlattices," in *Microcrystalline and Nanocrystalline Semiconductors. Symposium*, Symposium, 29 November 1994.) Pittsburgh, PA, USA: Material Research Society, p. 327, 1995.

[62]. Berger M.G., Thonissen M., Arens-Fischer R., Munder H., Luth H., Arntzen M., Theiss W., "Investigation and design of optical properties of porosity superlattices," *Thin Solid Films*, **255, 1-2**, p. 313, 1995.

[63]. Broadbent E.K., et al., *J. Vac. Sci.& Technol.*, **B17**, p. 2584, 1999.

[64]. Lehmann V., Ronnebeck S., "MEMS techniques applied to the fabrication of anti-scatter grids for X-ray imaging," *Sensors and Actuators* **A, 95**, p. 202, 2002.

INDEX